T0275821

LONDON MATHEMATICAL SOCIETY LECTURE NOTE SERIES

Managing Editor: Professor J.W.S. Cassels, Department of Pure Mathematics and Mathematical Statistics, University of Cambridge, 16 Mill Lane, Cambridge CB2 1SB, England

The books in the series listed below are available from booksellers, or, in case of difficulty, from Cambridge University Press.

LONDON MATHEMATICAL SOCIETY LECTURE NOTE SERIES

Managing Editor: Professor J.W.S. Cassels, Department of Pure Mathematics and Mathematical Statistics, University of Cambridge, 16 Mill Lane, Cambridge CB2 1SB, England

The books in the series listed below are available from booksellers, or, in case of difficulty, from Cambridge University Press.

London Mathematical Society Lecture Note Series. 181

Geometric Group Theory
Proceedings of the Symposium held in Sussex 1991

Volume 1

Edited by

Graham.A. Niblo
University of Southampton

and

Martin A. Roller
Universität Regensburg

CAMBRIDGE
UNIVERSITY PRESS

CAMBRIDGE UNIVERSITY PRESS
Cambridge, New York, Melbourne, Madrid, Cape Town, Singapore, São Paulo

Cambridge University Press
The Edinburgh Building, Cambridge CB2 2RU, UK

Published in the United States of America by Cambridge University Press, New York

www.cambridge.org
Information on this title: www.cambridge.org/9780521435291

© Cambridge University Press 1993

This book is in copyright. Subject to statutory exception
and to the provisions of relevant collective licensing agreements,
no reproduction of any part may take place without
the written permission of Cambridge University Press.

First published 1993

A catalogue record for this publication is available from the British Library

ISBN-13   978-0-521-43529-1   paperback
ISBN-10   0-521-43529-3   paperback

Transferred to digital printing 2005

Table of Contents

Table of Contents

Preface

In the summer of 1991 geometric group theorists gathered in Sussex for the Geometric Group Theory Symposium. For the first two weeks of July they met at the Workshop session in Sussex University, where they took part in informal research seminars. There were 35 talks, including a course on the theory of racks by Roger Fenn and a series of lectures on semi-hyperbolic groups by Martin Bridson. This culminated in a London Mathematical Society Spitalfields Lecture day, with talks by Alan Beardon, Walter Neumann and Hyman Bass.

During the third week the residential conference took place at the White House Conference Centre in Chelwood Gate. There was a total of 40 talks, including four lectures by Eliahu Rips on his classification of groups acting freely on $\mathbb{R}$-trees and two lectures by Andrew Casson on convergence groups and his proof of the Seifert conjecture.

We had asked the participants of the conference to survey their special field of interest. About half of the papers in the first volume of these proceedings are of this nature, and we hope this will provide a good overview of the state of the art of geometric group theory at the time of the conference. In particular, Michael Gromov's response to our request was so enthusiastic, that we decided to dedicate the entire second volume to his essay *Asymptotic Invariants of Infinite Groups*. The remainder of the articles are either extended versions of papers given at the conference or answers to questions posed there.

The first volume is concluded by a list of problems suggested at the Geometric Group Theory Symposium. Most were written up in the coffee lounge throughout the workshop and conference sessions, and were discussed at length by the Falmer Pub Research Group. The editors would like to express their thanks to Steve Gersten, who chaired the Problem Session at the conference with his customary good humour, and to Swarup, who faithfully recorded the discussions.

In editing these proceedings we have tried to produce some uniformity while respecting the authors' individuality. We of course accept full responsibility for any inconsistencies or errors introduced by our interference.

It is a pleasure to thank all the people who helped us in preparing and running the conference. Christine Coles assisted us enthusiastically from an early stage, David Epstein encouraged and advised us, and Martin Dunwoody and Swarup gave us valuable support during the conference. We would also like to thank the principal speakers whose enthusiasm, and willingness to commit themselves to the project at the beginning ensured its success.

The following provided generous financial support: Wolfram Research sponsored a demonstration of Mathematica; the London Mathematical Society funded the Spitalfields Lectures; the Science and Engineering Research Council, whose generous grant (Grant No. GR/G 35572) made the Symposium possible.

Graham A. Niblo
Mathematics Department,
Southampton University,
Southampton,
SO9 5NH
England

Martin A. Roller
Mathematik,
Universität Regensburg,
Postfach 397,
8400 Regensburg,
Germany

List of Participants

Abboud, E., Haifa
Alonso, J.M., Stockholm
Bacher, R., Geneve
Baik, Y.G., Glasgow
Bass, H., Columbia, New York
Benakli, N., Orsay
Bestvina, M., IHES
Bieri, R., Frankfurt
Bridson, M., Cornell, Ithaca
Brookes, C.J.B., Cambridge
Brunner, A., Wisconsin, Parkside
Burns, R.G., York Univ., Toronto
Casson, A., Berkeley
Chaltin, H., Marseille
Charney, R., Ohio State
Chiswell, I.M., QMW, London
Cho, J.R., Glasgow
Cohen, D.E., QMW, London
Collins, D.J., QMW, London
Devine, D., Sussex
Dicks, W., Barcelona
Dunwoody, M.J., Sussex
Edmonds, I., Sussex
Edjvet, M., Nottingham
Edwards, R.D., UCLA
Epstein, D.B.A., Warwick
Feighn, M.E., Rutgers, Newark
Fenn, R., Sussex
Geoghegan, R., Binghampton
Gersten, S., Utah
Gilbert, N., Edinburgh
Griffiths, D., King's College, London
Gromov, M., IHES
Haefliger, A., Geneve
de la Harpe, P., Geneve
Harvey, W., King's College, London
Holt, D.F., Warwick
Hotchkiss, P., SUNY Albany
Howie, J., Edinburgh
Huck, G., Northern Arizona
Humphries, S., Brigham Young
Imrich, W., Montan Univ., Leoben

Juhasz, A., Haifa
Kenyon, R., IHES
Kourouniotis, C., Crete
Kropholler, P.H., QMW, London
Leary, I., QMW, London
Long, D.D., Univ. of California
Luke, G., Oxford
Lustig, M., Bochum
Mikkelsen, M., Lyngby
Minsky, Y., Haifa
Moriah, Y., Haifa
Neumann, W., Ohio State
Niblo, G.A., QMW, London
Paton, M., QMW, London
Paulin, F., ENS Lyon
Peifer, D., Univ. of Illinois
Pittet, C., Geneve
Planche, P., Geneve
Potyagailo, L., Tel Aviv
Pride, S.J., Glasgow
Ranicki, A., Edinburgh
Raptis, E., Athens
Rees, S., Newcastle
Reid, A., Aberdeen
Rimlinger, F., Fairfield
Rips, E., Jerusalem
Roller, M., Sussex
Rosebrock, S., Frankfurt
Rourke, C., Warwick
Sela, Z., Jerusalem
Series, C., Warwick
Shapiro, M., Ohio State
Short, H., City Coll., New York
Simonsen, A., Lyngby
Stallings, J., Berkeley
Stoehr, R., Manchester
Swarup, G.A., Melbourne,
Talelli, O., Univ. of S. California
Thomas, R.M., Leicester
Ventura, E. , Barcelona
Vogtmann, K., Cornell Univ.
Yamasaki, M., Edinburgh

Group Actions and Riemann Surfaces

Alan F. Beardon

Department of Mathematics and Mathematical Statistics, University of Cambridge,
16 Mill Lane, Cambridge, CB2 1SB.

1. Introduction

We begin by considering two familiar ways of constructing Riemann surfaces.
First, we take a power series P converging on some disc D_0 with centre z_0, and
expand P about some point z_1 in D_0 other than z_0. In general, P will converge
in a disc D_1 extending beyond D_0, and if we continue this process indefinitely
we obtain a maximal Riemann surface on which the analytic continuation of
P is defined. Of course, if we return to a region where P is already defined,
but with different values, we create a new 'sheet' of the surface; thus we are
led to the notion of a Riemann surface constructed from a given power series:
this is the Weierstrass approach. A more modern approach is simply to define
a Riemann surface as a complex analytic manifold but either way, there is the
problem of showing that these two definitions are equivalent. It is easy enough
to see that a Riemann surface obtained by analytic continuation is an analytic
manifold, so we must focus our efforts on showing that every analytic manifold
supports an analytic function. One solution to this problem lies in showing
first that every such manifold arises as the quotient by a group action, and
second, that we can construct functions invariant under this group action.
As a by-product of a study of these groups we obtain important and very
detailed quantitative information about the geometric nature of the general
Riemann surface.

In pursuing this line of thought we are led naturally into the study of discrete group actions on the three classical geometries of constant curvature
(the sphere, the Euclidean plane and hyperbolic plane), and also to the way
that discreteness imposes severe geometric constraints on the corresponding
quotient surfaces. The crystallographic restriction on Euclidean groups is one
example of this, but, as we shall see, this idea reaches out well beyond the
study of wall-paper patterns and goes on to exert a powerful influence on

the geometric structure of all Riemann surfaces. It is this influence that we attempt to describe in this essay.

2. The Uniformisation Theorem

The Uniformisation Theorem implies that every Riemann surface can be realised as the quotient by a group action and, once we have given an explicit description of the possible groups that arise in this way, this lays bare the geometric nature of Riemann surfaces for all to see. In this section we discuss the group actions that arise and later we shall show how, by examining these, we can obtain universal information about the metric and the geometric structure of the general Riemann surface.

Consider now an arbitrary Riemann surface $\mathcal{R}$. It is a straightforward matter to construct the topological universal covering surface $\mathcal{S}$ and to lift the conformal structure from $\mathcal{R}$ to $\mathcal{S}$, thus realising $\mathcal{R}$ as the quotient of the simply connected Riemann surface $\mathcal{S}$ by the corresponding cover group G. We know (from topology) that G has certain interesting properties (for example, only the identity in G can have fixed points on $\mathcal{S}$, and G is the fundamental group of $\mathcal{R}$) but, so far, we have little concrete information about $\mathcal{S}$ or G.

The key step now is to invoke the strong form of the *Riemann Mapping Theorem,* namely that every simply connected Riemann surface is conformally equivalent to one of the three spaces

(a) the extended complex plane (or the Riemann sphere) $\mathbb{S}$,

(b) the complex plane $\mathbb{C}$,

(c) the hyperbolic plane $\mathbb{H}$ (that is, the unit disc $\{z : |z| < 1\}$ in $\mathbb{C}$).

With this available, we can pass (by a conformal mapping) from the universal cover $\mathcal{S}$ to one of these spaces and henceforth take one of $\mathbb{S}$, $\mathbb{C}$ or $\mathbb{H}$ as the universal cover of $\mathcal{R}$. We can immediately draw some interesting conclusions from this representation; for example, it shows that every Riemann surface has a countable base, and also an exhaustion by compact sets. (We remind the reader that there are surfaces which do not have these desirable properties.) The benefits of using only $\mathbb{S}$, $\mathbb{C}$ or $\mathbb{H}$ as universal covering spaces are enormous and these arise from the (easily proved) analytic fact that the group of conformal automorphisms of each of these spaces is a subgroup of the Möbius group $\mathcal{M}$ consisting of all maps of the form

$$g(z) = \frac{az + b}{cz + d}, \quad ad - bc = 1. \tag{1}$$

In fact, $\mathcal{M}$ is the full group of conformal automorphisms of $\mathbb{S}$, the conformal automorphisms of $\mathbb{C}$ are the maps $g(z) = az + b$, and conformal automorphisms of $\mathbb{H}$ are

$$g(z) = \frac{az + \bar{c}}{cz + \bar{a}}, \quad |a|^2 - |c|^2 = 1. \tag{2}$$

We have now reduced the study of the general Riemann surface (defined as a manifold) to the study of quotients by discrete subgroups of $\mathcal{M}$. Of course, each of the three spaces $\mathbb{S}$, $\mathbb{C}$ or $\mathbb{H}$ are metric spaces, the hyperbolic metric on $\mathbb{H}$ being derived from the line element

$$ds = \frac{2|dz|}{1 - |z|^2} \, ,$$

and we shall be particularly interested in the Möbius isometries of these spaces. In the case of $\mathbb{H}$, all automorphisms of the form (2) are hyperbolic isometries and these are the only conformal isometries of $(\mathbb{H}, \rho)$. For more information, see [2], [4], [5], [7], [11] and [12].

Each Riemann surface $\mathcal{R}$ determines its universal cover uniquely from among the three possibilities $\mathbb{S}$, $\mathbb{C}$ or $\mathbb{H}$ because these three spaces are conformally inequivalent. There are very few surfaces $\mathcal{R}$ which have $\mathbb{S}$ or $\mathbb{C}$ as their universal cover and we can easily dispose of these. If $\mathcal{R}$ has $\mathbb{S}$ as its universal cover, the cover group G is trivial (for only the identity in G can have fixed points on $\mathbb{S}$) and so (up to conformal equivalence) $\mathcal{R} = \mathbb{S}$. This shows that there is only one conformal structure on the sphere.

The case when $\mathcal{R}$ has $\mathbb{C}$ as its universal cover is hardly more interesting. As the automorphism group of $\mathbb{C}$ is the class of maps $z \mapsto az + b$, the requirement that only the identity in G can have fixed points on $\mathbb{C}$ means that G must be a group of translations (that is, $a = 1$). As G is also discrete, we see that either G is the trivial group (and $\mathcal{R} = \mathbb{C}$), or G is generated by one or two (linearly independent) translations. If G is a cyclic group then (up to conjugacy) G is generated by $z \mapsto z + 2\pi i$, the quotient map is $z \mapsto \exp z$ (for $\exp z = \exp w$ if and only if $w = g(z)$ for some g in G), and in this case $\mathcal{R}$ is the punctured plane $\mathbb{C} - \{0\}$. It is of interest to note that the surfaces $\mathbb{S}$, $\mathbb{C}$ and $\mathbb{C} - \{0\}$ which we have obtained so far are, collectively, the sphere with at most two punctures. In the remaining case, $\mathcal{R}$ has $\mathbb{C}$ as its universal cover, G is generated by two independent translations and $\mathcal{R}$ is a torus. Before moving on, we note that although any two tori are topologically equivalent, there are infinitely many distinct conformal equivalence classes of tori so, as far as the analyst is concerned, the discussion of this can go much further.

Our discussion so far has led to the conclusion that essentially every Riemann surface has the hyperbolic plane $\mathbb{H}$ as its universal cover and, as the local geometry of $\mathbb{H}$ projects down to $\mathcal{R}$, we see that *the intrinsic geometry of the generic Riemann surface is hyperbolic*; for example, if T is a small triangle on a generic Riemann surface then its angle sum is strictly less than π. This, of course, applies equally well to planar Riemann surfaces, so that when we study plane domains with the induced Euclidean structure in elementary complex analysis we are, in fact, using the wrong geometry. As a simple indication that we should have anticipated this, we observe that in any reasonable metric tied to a domain D, the boundary of D should be infinitely far from any point inside it: this is true in the intrinsic hyperbolic geometry of D but not, of course, generally true in the Euclidean metric. To be more precise, in any plane domain the hyperbolic metric is a Riemannian metric given by $ds = \lambda(z)|dz|$ where $\lambda(z) \to \infty$ as $z \to \partial D$. For 'most' domains, and for all simply connected domains, $\lambda(z)$ is of the same order as the reciprocal of the Euclidean distance of z from ∂D. There are, however, domains in which $\lambda(z) \to \infty$ at a slower rate than this as z approaches particular boundary points of D.

The realisation that the intrinsic geometry in complex analysis is hyperbolic has profound repercussions and we illustrate this here with one fundamental, but simple, result. One of the results in a standard complex analysis course is Schwarz's Lemma: if f is an analytic map of the unit disc into itself, and if $f(0) = 0$, then $|f(z)| \le |z|$ with equality if and only if f is a rotation. Now the hypotheses asserts that f is a self-map of the hyperbolic plane into itself, and the conclusion is stated (rather perversely) in terms of the Euclidean distance $|z|$ between z and 0. However, if ρ is the hyperbolic distance in $\mathbb{H}$ the conclusion can also be stated as

$$\rho(f(z), f(0)) \le \rho(z, 0)$$

and, with a little (but not much) extra work, we obtain the far more penetrating form known as the Schwarz-Pick Lemma: *if f is an analytic map of the hyperbolic plane into itself, then f is either a contraction or an isometry*, [1]. Many results in complex analysis depend as much on this fact as they do on analyticity.

It is now evident that we must examine discrete groups of hyperbolic isometries, and we recall that any conformal isometry of $\mathbb{H}$ is of the form

$$g(z) = \frac{az + \bar{c}}{cz + \bar{a}}, \quad |a|^2 - |c|^2 = 1.$$

A discrete group is necessarily countable, and the condition for discreteness is equivalent to the statement that $|a| \to \infty$ or, equivalently, to $|c| \to \infty$ as g runs through G. As $|a|^2 - |c|^2 = 1$, this is equivalent to $|a/c| \to 1$ and, as $|c/a| = |g(0)|$, we see that discreteness is equivalent to saying that all G-orbits accumulate only on the ideal boundary of $\mathbb{H}$ (that is, on the unit circle $|z| = 1$). We shall return to discuss the isometries in greater detail later.

We end this section with the remark that the Uniformisation Theorem is a deep result (as must be evident from the results that flow from it), and the most difficult part of the proof is the existence part, namely the Riemann Mapping Theorem. The usual proof of this involves potential theory (where the existence of the Green's function is the main issue) and we refer the reader to, for example, [1] and [3] for the details. It is, perhaps, worth mentioning that the existence or otherwise of a Green's function is a problem in real analysis.

3. Automorphic functions

If G is a group acting on a space X, then the class of maps $F : X/G \to Y$ corresponds precisely to the class of G-periodic maps $f : X \to Y$, that is, to maps f with the property that $f(gx) = f(x)$ for every x in X and every g in G. As an example, each analytic map $f : \mathbb{C} \to \mathbb{C}$ with period 2π corresponds to an analytic map $F : \mathbb{C} - \{0\} \to \mathbb{C}$ which has a Laurent expansion

$$F(\zeta) = \sum_{n=-\infty}^{+\infty} a_n \zeta^n$$

valid throughout $\mathbb{C} - \{0\}$. Here, the cover group G is generated by $z \mapsto z + 2\pi$, the quotient map is $z \mapsto e^{iz}$ and so the periodic map f has a *Fourier expansion*

$$f(x + iy) = F(e^{iz}) = \sum_{n=-\infty}^{+\infty} a_n e^{inx - ny}$$

valid throughout $\mathbb{C}$ (this reduces to the usual Fourier series when $y = 0$). For more details, see [8], [9] and [10]. As an illustration of this, $z \mapsto \cos z$ corresponds to $F(\zeta) = (\zeta + \zeta^{-1})/2$. As F is a rational map of degree 2, this explains why cos is a 2-1 mapping of each fundamental strip (of width 2π) onto the sphere $\mathbb{S}$.

One of the consequences of the Uniformisation Theorem is that given any Riemann surface $\mathcal{R}$ (as a manifold), we can construct analytic functions defined on it simply by constructing analytic G-periodic functions on the appropriate covering space, and hence we can establish to equivalence of the two concepts discussed in Section 1. Of course, any function of the form $\sum_{g \in G} f(g(z))$ is periodic providing that the sum is absolutely convergent (or G is finite); however, in general the sum will diverge and some modifications are required to force convergence.

In the case of the Weierstrass elliptic function $\wp$, for example, (where $\mathcal{R}$ is a torus) the group G is generated by two independent translations $z \mapsto z + l$ and $z \mapsto z + \mu$ and one defines

$$\wp(z) = \frac{1}{z^2} + \sum_{\substack{m,n \in \mathbb{Z} \\ (m,n) \neq (0,0)}} \left(\frac{1}{(ml + n\mu - z)^2} - \frac{1}{(ml + n\mu)^2} \right).$$

Without the term independent of z, the series would diverge and the subtraction of this term is designed to make the difference small enough to force absolute convergence (as indeed it does).

Our interest centres on the generic Riemann surface with the hyperbolic plane
$\mathbb{H}$ as its covering space, so our task is really to take any discrete group G of
hyperbolic isometries acting on $\mathbb{H}$, and to learn how to construct periodic
functions or, as they are more commonly known in the subject, *automorphic
functions*. The key step is to note that, ignoring questions of convergence for
the moment, if

$$\theta(z) = \sum_{g \in G} f(g(z))g'(z)^t, \tag{3}$$

then, for h in G,

$$\theta(h(z))h'(z)^t = \sum_{g \in G} f(gh(z))g'(hz)^t h'(z)^t = \sum_{\gamma \in G} f(\gamma(z))\gamma'(z)^t = \theta(z),$$

and so the quotient of any two such θ-functions (constructed using two different choices of f) will be automorphic. In fact, any function θ here gives
rise to a differential on the Riemann surface: for more details, see [5], [8], [9]
and [10].

We must now pay attention to the convergence. First, we impose reasonable
properties on the function f in (3) (which, in this exposition, we ignore), and
then attempt to obtain absolute convergence by studying the series

$$\sum_{g \in G} g'(z)^t. \tag{4}$$

Now the general isometry is given by g in (2) and

$$g'(z) = \frac{1}{(cz + \bar{a})^2} = \frac{1}{c^2(z - g^{-1}(\infty))^2}.$$

The point $g^{-1}(\infty)$ lies outside the unit circle so, if z lies in some compact
subset K of $\mathbb{H}$, we obtain uniform convergence of (4) on K providing that the
series $\sum |c|^{-2t}$ converges.

The convergence of the series $\sum |c|^{-2t}$ for sufficiently large values of t is easily
established. For example, as $|g'(z)|^2$ represents the distortion in the *Euclidean*
area of the mapping g, we can take a small disc D in $\mathbb{H}$ in a fundamental
domain for G and then the Euclidean area of $g(D)$ is approximately $|g'(z)|^2 \mathcal{A}$,
where $\mathcal{A}$ is the Euclidean area of D. As the sets $g(D)$, $g \in G$, are disjoint
sets contained in $\mathbb{H}$ (of finite Euclidean area π), the series $\sum |c|^{-4}$ converges.
This argument can be made precise without difficulty, but it is natural to
object to it on the grounds that it invokes Euclidean arguments. It can be
modified to use hyperbolic arguments (although one should note that $\mathbb{H}$ has
infinite hyperbolic area) and in this way one can show that $\sum |c|^{-2t}$ converges
for all $t > 1$. In fact, this is the best one can do, for there are groups
for which the series diverges when $t = 1$ (the groups whose fundamental
region has finite area, and which correspond to compact Riemann surfaces

with finitely many punctures). For a given group G, one can examine the exponent of convergence of the series for that group (essentially, the smallest t for which the series (4) converges); for most groups this is the Hausdorff dimension of the limit set (the accumulation points of the orbits of G) and it is an important measure of the (common) rate at which the G-orbits move out to the ideal boundary of $\mathbb{H}$.

4. The geometry of Riemann surfaces

Before we turn our attention to cover groups of hyperbolic Riemann surfaces, let us briefly consider the ideas involved by examining the effect of discreteness on Euclidean groups. Suppose that Γ is a discrete group of Euclidean isometries acting on the Euclidean plane, and that Γ contains translations. By discreteness, the translations in Γ have a minimal translation length which we may assume is 1 and attained by the translation t in Γ.

Let g in Γ be a rotation of angle $2\pi/n$ about the point z_g, so that the conjugate element $h = tgt^{-1}$ is a rotation of the same angle about $t(z_g)$. Now draw a line L_2 through z_g and $t(z_g)$, a line L_1 through z_g making an angle π/n with L_2, and a line L_3 parallel to L_1 through $t(z_g)$. Let α_j denote reflection in L_j. Then g (or g^{-1}) is $\alpha_1\alpha_2$, h^{-1} (or h) is $\alpha_2\alpha_3$, and gh (or a similiar word) is the translation $\alpha_1\alpha_3$ through a distance $2|z_g - t(z_g)|\sin(\pi/n)$. As this is at least 1, and as $|z_g - t(z_g)| = 1$, we conclude that $n \le 6$.

This is the familiar crystallographic restriction, but the essential point here is to realise that these techniques are available in other geometries although the conclusions are different. The conclusions are different, of course, because each geometry comes equipped with its own trigonometry and the quantitative results of this type must, of necessity, reflect that particular trigonometry. To see the effect that the different trigonometries have on the three geometries, observe that Pythagoras' Theorem for a right angled triangle with sides a and b and hypothenuse c is $a^2 + b^2 = c^2$ in the Euclidean plane, whereas in the hyperbolic plane it is

$$\cosh a \cosh b = \cosh c,$$

and in the spherical case,

$$\cos a \cos b = \cos c.$$

It is only in the spherical case that we can have $a = b = c$. Note also that in the hyperbolic plane, for very large triangles we have (essentially) $a + b = c + \log 2$, that is, the vertices appear to be almost collinear (this is the effect of negative curvature). Again, in the hyperbolic plane there are regular n-gons with all angles $\pi/2$ precisely when $n \ge 5$; the case $n = 4$ is Euclidean and $n = 3$ is spherical.

In the hyperbolic plane the circumference of a circle grows exponentially with the radius, and roughly at the same rate as the area, so that we may regard the hyperbolic plane as having, by comparison with the Euclidean plane, an immense amount of room near its ideal boundary (the circle at infinity): formally, this is the effect of negative curvature and separating geodesics and *it is this that accounts for the inexhaustible supply of discrete hyperbolic groups compared with the* 17 *wallpaper groups in the Euclidean case*. Roughly speaking, we can construct an increasing sequence of discrete hyperbolic groups in which in each case there is always ample room in the space near infinity to incorporate some extra group action into the picture.

In order to understand the mechanism by which discreteness imposes geometric constraints on hyperbolic Riemann surfaces, we need to know that an individual hyperbolic isometry g can be expressed as the composition $g = \alpha\beta$ of two reflections α and β across hyperbolic geodesics ℓ_α and ℓ_β, respectively (called the *axes* of α and β). We omit the proof of this, but remark that it is merely the hyperbolic counterpart of the familiar Euclidean fact that the composition of two reflections is either a rotation (if the axes of the reflection meet) or a translation (if the axes are parallel). It is a consequence of the failure of the Parallel Axiom in hyperbolic geometry that there are *three* possibilities in the hyperbolic plane, for there two geodesics either cross or not and, if not, they may or may not meet on the circle at infinity.

If ℓ_α and ℓ_β have a common orthogonal geodesic L_g; then g leaves L_g invariant. In fact, one end-point of L_g is an attracting fixed point, the other end-point is a repelling fixed point, and g moves each point of L_g by the same distance (twice the hyperbolic distance between ℓ_α and ℓ_β) along L_g. We call L_g the *axis* of g, the distance g moves each point along L_g is the *translation length* T_g of g, and g is said to be a *hyperbolic translation* (or, sometimes, a *loxodromic* isometry or even, rather confusingly, a hyperbolic element of the isometry group). In any event, if g lies in some cover group, the lines ℓ_α and ℓ_β cannot cross (else g would fix the point of intersection) and the only other case is that the two lines meet on the circle at infinity; in this case, g is said to be *parabolic* and it is conjugate to a Euclidean translation. In the case of a hyperbolic translation g, there is a useful formula for the distance a point z is moved by g, namely

$$\sinh \tfrac{1}{2}\rho(z, gz) = \sinh(\tfrac{1}{2}T_g)\cosh \rho(z, L_g);$$

thus the minimum movement occurs on the axis L_g of g, and the further z is away from the axis, the more it is moved by g.

Now consider two hyperbolic translations g and h such that the axes L_g and L_h have a common orthogonal geodesic ℓ_β (and are therefore necessarily disjoint). It is easy to see that these can be expressed in the form $g = \alpha\beta$ and $h = \beta\gamma$, where α, β and γ are reflections with axes ℓ_α, ℓ_β and ℓ_γ, respectively, and, as a consequence of this, we have $gh = \alpha\gamma$; in other words, the composition for gh is expressed neatly in terms of the compositions for g and h. Sadly, this argument is restricted to two dimensions for in higher dimensions g and h may each be a composition of more that two reflections and then cancellation (of β, for example) may not occur.

We now examine some of the consequences of this expression for gh. If L_g and L_h come close to each other compared with T_g and T_h, and if the lines ℓ_α and ℓ_γ are on the same side of ℓ_β (which may be achieved by replacing g by g^{-1} if necessary), then the lines ℓ_α and ℓ_γ will meet and gh will have a fixed point in $\mathbb{H}$. The reader is urged to draw a diagram to illustrate this but, roughly speaking, the lines L_g and L_h curve (in Euclidean terms) away from each other, and this forces the lines ℓ_α and ℓ_γ to cross providing that T_g and T_h are not too large. If g and h are in some cover group then gh cannot have fixed points in $\mathbb{H}$ and so, in any cover group, the geometric quantities T_g, T_h and $\rho(L_g, L_h)$ (that is, the hyperbolic distance between the axes of g and h) cannot all be small. As the axes of g and h project to closed geodesic loops on the Riemann surface $\mathcal{R}$, this result tells us that *two short disjoint geodesic loops on $\mathcal{R}$ must be fairly far apart* or, equivalently, if they are close to each other then one is fairly long. The same argument applies when g and h are in the same conjugacy class in G, and this implies that if the distance from a geodesic loop to itself along a non-trivial closed curve on $\mathcal{R}$ is small, then the loop must be long. There are many results of this type available, and they all have a precise, quantitative, formulation which can be derived from elementary hyperbolic trigonometry; for example, the result above is that, in all cases,

$$2\sinh(\tfrac{1}{2}T_g)\sinh(\tfrac{1}{2}T_h)\sinh^2\tfrac{1}{2}\rho(L_g, L_h) \geq 1. \tag{5}$$

Roughly speaking, these results describe the universal metric properties of handles on a hyperbolic Riemann surface.

There is an analogous inequality to (5), valid when the two axes cross at an angle θ, and this is

$$\sinh(\tfrac{1}{2}T_g)\sinh(\tfrac{1}{2}T_h)|\sin\theta| \geq 1. \tag{6}$$

Suppose now that σ is a self-intersecting loop on a Riemann surface $\mathcal{R}$ of length $|\sigma|$. Then, in terms of the cover group G, this means that there is a hyperbolic translation g and a conjugate element hgh^{-1} such that both have translation length $|\sigma|$, and such that their axes cross at some angle θ. Applying (6), and using the fact that $1 \geq |\sin\theta|$, we see that $\sinh(\tfrac{1}{2}|\sigma|) \geq 1$; thus there is a lower bound on the length of a non-simple loop on a hyperbolic Riemann surface, and this is a universal lower bound in the sense that *it does not depend on the particular surface*.

A related result, quite beautiful in its simplicity, is that if G is a group of hyperbolic isometries without hyperbolic rotations, then, for all z in $\mathbb{H}$,

$$\sinh\tfrac{1}{2}\rho(z, gz)\sinh\tfrac{1}{2}\rho(z, hz) \geq 1 \tag{7}$$

unless g and h lie in some cyclic subgroup of G (see [2]). In this result, G is *not* assumed to be discrete, and it is a consequence of this that *any group of hyperbolic motions without rotations is automatically discrete*, and hence automatically a cover group of some surface. Equally, if G is a cover group, this result is still applicable and it implies (but in a rather stronger form) that the lengths of any two loops from any point in the surface $\mathcal{R}$ can only both be small when they lie in a cyclic subgroup of the homotopy group. Again, this is a quantitative universal result which holds for all hyperbolic Riemann surfaces. Of course, (7) need not hold if $\langle g, h \rangle$ is cyclic; for example, one can construct a small translation g and then put $h = g^2$.

Results of this type are not restricted to hyperbolic translations, and there are similiar conclusions to be drawn for all elements in a cover group G. Briefly, a puncture p on the surface $\mathcal{R}$ corresponds to the unique fixed point ζ of a parabolic element g of G and, conversely, any parabolic element determines a puncture on $\mathcal{R}$ (we recall that a parabolic element is conjugate to the Euclidean translation $z \mapsto z + 1$ acting on the upper half-plane model of $\mathbb{H}$).

It can be shown that each such parabolic fixed point ζ is the point of tangency of a horocycle cal H_g in $\mathbb{H}$ (cal H_g is a Euclidean disc in $\mathbb{H}$ tangent to $\{z : |z| = 1\}$ at ζ) and, moreover, the horocycles cal H_g can be chosen to be disjoint for distinct ζ and to be compatible with conjugation in the sense that $h(\text{cal}\,H_g) = \text{cal}\,H_{hgh^{-1}}$. In fact, the elements of G leaving cal H_g invariant form a cyclic subgroup G_0 of G and the quotient space cal H_g/G_0 is a once punctured disc which is conformally equivalent to a neighbourhood of the corresponding puncture p on $\mathcal{R}$. This is the formal statement of the fact that the neighbourhood of any puncture on any hyperbolic Riemann surface is the same as the quotient of $\{x + iy : y > 0\}$ by the exponential map. As far as geometric constraints are concerned, one can show that the cal H_g can be chosen so that the (finite) area of cal H_g/G_0 is at least 1 (a universal lower bound), and that in an appropriate and universal sense the simple geodesic loops on the surface $\mathcal{R}$ do not come too close to the puncture p. The geometry near a puncture p is such that one travels an infinite distance to reach p, but through a neighbourhood of p of finite area; thus, roughly speaking, the surface $\mathcal{R}$ has an infinitely tall, infinitely thin, spike at p.

It is natural to now allow our groups to contain hyperbolic rotations and so study the most general discrete group G acting on the hyperbolic plane: this amounts to considering branched coverings of Riemann surfaces. In fact, if G is finitely generated, then it contains a torsion-free normal subgroup of finite index (this is *Selberg's Lemma*) and so we may expect that suitably, but only slightly, relaxed versions of the earlier constraints will hold. This is so but, of course, there are also entirely new types of constraints to be considered, namely those involving rotations.

Briefly, we describe some of the many geometric constraints that hold for these groups. Suppose, for simplicity, that G is finitely generated. Then G has a fundamental region $\mathcal{F}$ consisting of a finite number of sides (and also many other desirable features) and we mention, in passing, that a presentation of G can be read off from the geometry of $\mathcal{F}$; in particular, the relations in G are consequences of the relations associated with small loops around each vertex of $\mathcal{F}$. If $\mathcal{F}$ has finite area $\mathcal{A}$ (so that the corresponding Riemann surface $\mathcal{R}$ is compact apart from a finite number of punctures), we can compute $\mathcal{A}$ from the topological and geometric data using the Gauss-Bonnet Theorem. The formula for $\mathcal{A}$ involves the genus γ of $\mathcal{R}$, the number k of punctures of $\mathcal{R}$, and the orders m_j of the finite number of (conjugacy classes) of rotation subgroups of G, and is

$$\mathcal{A}/2\pi = 2\gamma - 2 + k + \sum_{j=1}^{r} \left(1 - \frac{1}{m_j}\right).$$

It is not hard to see that the function (of integer variables) on the right hand side attains its positive minimum value (for the so-called $(2,3,7)$-Triangle group) and that this minimum is $1/42$. Although this shows that every Riemann surface has area of at least $\pi/21$, there are no rotations in a cover group and so one one can do better; the finite sum in the formula for $\mathcal{A}$ is absent, the area is at least 2π and again this is a universal inequality.

Suppose now that g and h are rotations in G of orders p and q, respectively, with fixed points z_g and z_h (where $p \geq 3$ and $q \geq 3$, although one, but not both, of p and q can be 2). Clearly, any fundamental region can be assumed to lie in a cone-shaped region with vertex z_g and angle $2\pi/p$ (a fundamental region for $\langle g \rangle$), and likewise for h. The intersection of these two cones will contain a fundamental region for G and so have area at least $\pi/21$; thus it is evident that, given p and q, the hyperbolic distance $\rho(z_g, z_h)$ cannot be too small. There is a class of groups known as the Triangle groups (these are the groups whose quotient has small area, and for which the orbits are the most closely packed in $\mathbb{H}$) for which the argument is a little fussy and the results not quite as strong, but, excluding these, a careful geometric argument based on this idea and a calculation of the area involved leads to the sharp result

$$\cosh \rho(z_g, z_h) \geq \frac{1 + \cos(\pi/p)\cos(\pi/q)}{\sin(\pi/p)\sin(\pi/p)}.$$

Briefly, the distance between a fixed point of order p and a fixed point of order q (possibly $p = q$ here) is bounded below by a universal function of p and q.

As the previous inequality suggests, it is true that there is no positive lower bound on the distance between two fixed points of order two. However, there is a lower bound on the triangle bounded by three non-collinear fixed points of order two. Indeed, the composition of two rotations of order two produces a hyperbolic translation whose translation length is twice the distance between the fixed points; hence if three fixed points are close and non-collinear, we have two translations with crossing axes and small translation length and this violates discreteness. This constraint can also be expressed in a sharp quantitative form.

5. Actions of discrete Möbius groups

We turn now to a discussion of what can be said about the most general discrete subgroup G of the Möbius group. First, we must distinguish carefully between the *discreteness* of G (as a topological group) and its *discontinuous action* (G acts discontinuously in a domain D if, for each compact subset K of D, $g(K) \cap K \neq \emptyset$ for only a finite set of g in G). It is easy to see that if G acts discontinuously in some D then G is discrete, but, as the following example shows, the converse is false.

The Picard group Γ is the group of Möbius maps g of the form (1) where the coefficients a, b, c and d are Gaussian integers. Clearly, Γ is discrete. However, for each Gaussian integer ζ and each (real) integer N, the map

$$h(z) = \frac{(1 - N\zeta)z + \zeta^2}{-N^2 z + (1 + N\zeta)}$$

is in Γ and fixes ζ/N. As the set of points ζ/N is dense in $\mathbb{C}$, Γ does not act discontinuously in any open subset of $\mathbb{C}$.

In the light of this example, we need another way of looking at discrete subgroups of $\mathcal{M}$ and this is possible if we pass to three dimensional hyperbolic space in the following way. We embed the complex plane $\mathbb{C}$ naturally in $\mathbb{R}^3$ as the plane $x_3 = 0$ and view each circle in $\mathbb{C}$ as the equator of a sphere in $\mathbb{R}^3$, and each line in $\mathbb{C}$ as the intersection of a vertical plane with $\mathbb{C}$. Obviously, each reflection α in a line, or circle, in $\mathbb{C}$ extends to the reflection in the corresponding spheres or plane and we do not distinguish between the original α and its extension. Now each Möbius map can be expressed as the composition of at most four reflections (in circles or straight lines in $\mathbb{C}$) and so, in the obvious way, each Möbius map g extends to an action on $\mathbb{R}^3$ and $\mathbb{C}$ is invariant under this action. We call this the Poincaré extension of g.

The important feature of the extension described above is that the upper-half of $\mathbb{R}^3$ with the Riemannian metric $ds = |dx|/x_3$ is a model of hyperbolic 3-space $\mathbb{H}^3$ and each Möbius map g now *acts as an isometry of this space*. From a more general point of view, we now recognise that the true domain of the Möbius maps (1) is $\mathbb{H}^3$ (where they act as isometries) and *not* the extended complex plane (the boundary of $\mathbb{H}^3$) where they are always introduced. The relationship between discreteness and discontinuity now reappears, for a general subgroup of $\mathcal{M}$ is discrete (as a topological group) *if and only if* it acts discontinuously in $\mathbb{H}^3$ (this is the higher dimensional version of the result that a subgroup of $SL(2, \mathbb{R})$ is discrete if and only if it acts discontinuously on the upper half of the complex plane). In fact, this result can be put into a sharp quantitative form: if g is given by (1), we write

$$||g||^2 = |a|^2 + |b|^2 + |c|^2 + |d|^2,$$

so that $||g||$ is the usual matrix norm, and then we have

$$||g||^2 = 2 \cosh \rho(j, g(j)),$$

where $j = (0, 0, 1)$. This relation shows that as g runs through G, $||g|| \to \infty$ (that is, G is discrete) if and only if $\rho(j, g(j)) \to \infty$ (that is, G acts discontinuously in upper half-space). In addition, it also shows that g fixes the point j if and only if $||g||^2 = 2$, equivalently, if and only if the matrix for g is in $SU(2, \mathbb{C})$. Of course, as the Möbius group acts transitively on $\mathbb{H}^3$, the stabiliser of any point is conjugate to this.

We can even describe the action of Möbius maps on $\mathbb{H}^3$ algebraically. We view the upper-half $\mathbb{H}^3$ of $\mathbb{R}^3$ as the set of quaternions $\zeta = x + iy + jt + 0k$ and then let the map g given by (1) act on $\mathbb{H}^3$ by the rule

$$\zeta \mapsto (a\zeta + b).(c\zeta + d)^{-1}, \tag{8}$$

the computations being carried out in the space $\mathcal{Q}$ of quaternions. Of course, great care must be taken in calculations as multiplication of quaternions is not commutative, but this does indeed give the correct action of g. Using the quaternion representation, we see that if $g(j) = j$ then

$$j = g(j) = (aj + b)(cj + d)^{-1} = \frac{(b\bar{d} + a\bar{c}) + j}{|c|^2 + |d|^2},$$

or, equivalently,

$$b\bar{d} + a\bar{c} = 0, \quad |c|^2 + |d|^2 = 1.$$

If we now appeal to the identity

$$|b\bar{d} + a\bar{c}|^2 + 1 = |b\bar{d} + a\bar{c}|^2 + |ad - bc|^2 = (|a|^2 + |b|^2)(|c|^2 + |d|^2),$$

this gives $||g||^2 = 2$. It is of interest to note that $\mathbb{R}^3$ is embedded in $\mathcal{Q}$ as the set of quaternions with zero k-component and the action of g given by (8) actually preserves this property. It is not at all obvious why this happens but the explanation becomes clearer when we pass to higher dimensions.

The processes we have just described extend without difficulty to all dimensions, and each Möbius map acting in $\mathbb{R}^n$ (as a composition of reflections in $(n-1)$-planes and spheres) extends to an isometry of $\mathbb{H}^{n+1}$. If C_n denotes the Clifford algebra generated by $1, i_1, \ldots, i_n$ with $i_j^2 = -1$ and $i_j i_k = -i_k i_j$ when $j \neq k$, then the maps

$$g(\zeta) = (a\zeta + b).(c\zeta + d)^{-1}, \tag{9}$$

with suitable, but mild, restrictions on the coefficients, act on, and preserve, the linear subspace V of C_n spanned by the generators $1, i_1, \ldots, i_n$. The elements of V are linear combinations of $1, i_1, \ldots, i_n$ and we call such points the *vectors* of C_n. The elements of C_n that are products of non-zero vectors form a group (the Clifford group) and the restrictions on the coefficients in (9) are that they are in the Clifford group and that they satisfy a type of determinental condition (which corresponds to $ad - bc = 1$ in the complex case). The computations here are not always as productive as one might hope, largely, perhaps, because the algebra is less helpful; for example,

$$(i_1 + i_2 i_3 i_4)^2 = 0$$

so that C_n contains divisors of zero. It is of interest to note that the field of complex numbers is C_1 and it is only in this (low dimensional) case that the entire Clifford algebra coincides with its subspace of vectors: in this sense, the familiar complex Möbius maps are a special case indeed.

Let us illustrate these ideas in the context of $\mathbb{R}^3$. In this case we are working in C_2, which is generated by $1, i_1 (= i)$, and $i_2 (= j)$, and the map (9) preserves the space V of vectors $x + yi_1 + ti_2 + 0i_1 i_2$, where, of course, $i_1 i_2 = ij = k$. As might be expected, the familiar representation of rotations of Euclidean 3-space by quaternion maps is a part of this much larger picture. To be more explicit, the rotations of the ball in $\mathbb{R}^3$ are (after stereographic projection) represented by the Möbius maps

$$g(z) = \frac{az + b}{-\bar{b}z + \bar{a}}, \quad |a|^2 = |b|^2 = 1,$$

and the matrix for g is one form for a quarternion of unit norm.

Returning now to the geometry, the action of the Möbius group on upper half-space $\mathbb{H}^n$ can be transferred (by a conjugacy in the full Möbius group acting in $\mathbb{R}^n$) to the unit ball B^n and in this way we recapture the ball model of hyperbolic n-space complete with the appropriate Möbius maps as its isometry group. The Brouwer Fixed Point Theorem (for example) guarantees that our isometries have fixed points in the closure of B^n and a more careful examination shows that each isometry either has one or two fixed points on ∂B^n, or it has an axis of fixed points in B^n. We call elements of the latter type *elliptic elements*: for example, the Euclidean rotation $z \mapsto iz$ in $\mathbb{C}$ extends to $\mathbb{H}^3$ as a rotation of order 4 which fixes each point of the positive x_3 axis.

Let us now return to the original discussion of a general discrete group G of Möbius maps of the form (1) acting on the extended complex plane and deal first with two special cases. One can show that if such a group G contains *only* elliptic elements, then the elements of G have a common fixed point in $\mathbb{H}^3$. This means that, up to conjugation, and using the unit ball model, G is a discrete subgroup of the compact group of rotations of $\mathbb{R}^3$ and so is finite. With a little more work, one can go on to show that G must be the symmetry group of one of the five regular Platonic solids.

For the second special case, suppose that G is a discrete subgroup of automorphisms of $\mathbb{C}$ (that is, of maps of the form $z \mapsto az + b$). Then every element of G fixes ∞ and the crucial (but elementary) point to observe now is that if both a translation and a dilatation in G fix ∞, then G cannot be discrete. As a consequence, G cannot contain both translations and dilatations, and this is the key step in analysing the possibilities for G in this case. With this, we can show that G is either a finite cyclic rotation group, or a finite extension of a cyclic group generated by a dilatation, or a frieze group (when the translations in G are in the same direction) or, finally, one of the seventeen 'wallpaper groups'. To discuss the most general discrete group of (complex) Möbius maps, we can pass into $\mathbb{H}^3$ and study its action there. The group acts as a discontinuous group of isometries, and we can construct a fundamental region (a hyperbolic polyhedron) for the action of G and study the geometry of this. Most of the ideas used in the two-dimensional study are available and many (but not all) of the results remain true in this and the higher dimensional situations. There is, however, the problem mentioned above. The geometry of the hyperbolic plane, and the decomposition of the isometries into a product of two reflections is infinitely simpler than in higher dimensions. In $\mathbb{H}^3$, for example, the general isometry (that is, the general Möbius map) can be written as a composition of at most *four* reflections across hyperbolic planes and, in general, we need all four reflections. This fact alone implies that there is no longer a convenient way to represent the composition of two isometries as was the case in $\mathbb{H}^2$, and we must now turn to algebraic methods to proceed.

It is helpful, therefore, to now regard Möbius maps as elements of the matrix group $SL(2, \mathbb{C})$ and to consider conditions which must be satisfied by the matrix elements of f and g in order that $\langle f, g \rangle$ is discrete. The first significant result of this kind appeared in the 1930's or so and is simply this: if

$$f = \begin{pmatrix} 1 & 1 \\ 0 & 1 \end{pmatrix}, \quad g = \begin{pmatrix} a & b \\ c & d \end{pmatrix}, \tag{10}$$

and if $\langle f, g \rangle$ is discrete then

$$c = 0 \quad \text{or} \quad |c| > 1. \tag{11}$$

When applied to isometries of $\mathbb{H}^2$, this is simply the statement that the invariant horocycle based at a parabolic fixpoint has area at least 1, and it can be given a similiar interpretation in the geometry of $\mathbb{H}^3$. In higher dimensions, however, the nature of parabolic elements changes (and one obtains screw motions, for example) and the result generalises to only a limited extent.

To prove (11), one uses the technique which apparently goes back to Jordan, namely, starting with two given elements f and g one constructs iteratively a sequence h_n of commutators (or sometimes conjugates). If one of the original elements is near enough to the identity the sequence h_n converges to the identity and so, if the group is discrete, $h_n = I$ for all sufficiently large n. One must then convert $h_n = I$ into specific information about f and g and, excluding these case, we find that for discreteness, neither f nor g can be too close to I. This technique was exploited by Jorgensen around 1972, who used it to show that if the elements f and g in $SL(2,\mathbb{C})$ generate a non-elementary discrete group (roughly, elementary means a common fixed point here), then

$$|\text{trace}^2(f) - 4| + |\text{trace}(fgf^{-1}g^{-1})| \geq 1. \tag{12}$$

This is now known as *Jorgensen's inequality* and taking f and g as in (10), we recapture (11).

In 1979 Brooks and Matelski noticed that the iterative process used in the proofs of (11) and (12) is *quadratic*, and hence is exactly of the type studied in the Julia-Fatou theory of iteration of complex quadratic maps (now usually written as $z \mapsto z^2 + c$ and more popularly associated with the striking pictures of the Mandelbrot and Julia sets), and indeed, they produced, in this context, the first (albeit crude) picture of the Mandelbrot set. More recently, Gehring and Martin have refined and extended these ideas and have produced many more results of this type. With care, some of these ideas also extend to higher dimensions and other spaces.

6. Analytic functions again

We end this essay by returning to our point of entry, namely analytic funtions, to discuss briefly the impact our discussions have on complex analysis. We have already mentioned the Schwarz Lemma (arguably the most important result in geometric function theory), but the implications go far beyond this. Roughly speaking, geometric information on a group action, or a Riemann surface, can usually be converted into statements about analytic functions and the vehicle by which the information is transmitted is the associated hyperbolic (Riemannian) metric. As a single example of this, we quote *Landau's Theorem* (see [6]): there is a positive number k such that if

$$f(z) = \sum_{n=0}^{\infty} a_n z^n$$

is analytic in the unit disc $\mathbb{H}$ and omits the values 0, 1 and (of course) ∞, then

$$|a_1| \leq 2|a_0|(|\log|a_0|| + k).$$

This inequality, which controls the distortion due to f, is satisfied by *all* such functions (that is, k does not depend on f), and it is illuminating to examine in detail the role of k and the triple $\{0, 1, \infty\}$ in this.

First, we have seen that the sphere with one or two points removed has the complex plane as its universal cover. The only other case that arose with $\mathbb{C}$ as universal cover is the torus, thus a plane domain has $\mathbb{H}$ as its universal cover if and only if it its complement on the sphere contains at least three points. The actual value of these three points is irrelevant (for any triple can be mapped onto any other triple by a Möbius map); thus the significance of the triple of excluded values in Landau's Theorem is simply that f maps $\mathbb{H}$ *into some hyperbolic domain*. The significance of k is much more subtle ; in fact, if the hyperbolic metric on the complement Ω of $\{0, 1, \infty\}$ is $l(z)|dz|$, then $k = l(-1)$, a hyperbolic quantity. The moral of all this is clear, namely that complex analysis is intimately and inextricably linked with hyperbolic geometry. Such penetrating insights are often hard to establish and usually can only be seen retrospectively, but the effort involved is well worthwhile.

Finally, we mention Picard's Theorem, the proof of which is a supreme example of the strength of the methods outlined in this essay and is almost unbelievably short. *Picard's Theorem* states that if f is analytic on $\mathbb{C}$ and omits two values there, then f is constant. Let Ω be the complex sphere with the points 0, 1 and ∞ removed. The universal cover of Ω is the unit disc $\mathbb{H}$ and so (using the properties of the covering map) f lifts to a map F of $\mathbb{C}$ into $\mathbb{H}$. By Liouville's Theorem F is constant, thus so is f and the proof is complete.

7. The epilogue

This essay is an attempt to show the development of the geometry of Möbius group actions and Riemann surfaces towards its current state which, as many different aspects of mathematics seem to be merging together, now appears to be of interest to a much wider audience. It does not contain a catalogue of the latest results, nor does it attempt to cover all points of view within the subject, but hopefully, it does provide, in a fairly leisurely way, some geometric insight into Möbius group actions. For general reading, we refer the interested reader to any of the texts listed below.

References

[1]   L.V. Ahlfors, *Conformal invariants in geometric function theory*, McGraw-Hill, 1973.

[2]   A.F. Beardon, *The geometry of discrete groups*, Graduate Texts in Mathematics 91, Springer, 1983.

[3]   A.F. Beardon, *A primer on Riemann surfaces*, London Math. Soc. Lecture Notes 78, Cambridge University Press, 1984.

[4]   W. Fenchel, *Elementary geometry in hyperbolic space*, Walter de Gruyter, 1989.

[5]   W.J. Harvey, *Discrete groups and automorphic functions*, Academic Press, 1977.

[6]   W.K. Hayman, *Meromorphic functions*, Oxford Univ. Press, 1964.

[7]   G.A. Jones and D. Singerman, *Complex functions*, Cambridge Univ. Press, 1987.

[8]   I. Kra, *Automorphic forms and Kleinian groups*, Benjamin, 1972.

[9]   J. Lehner, *A short course on automorphic functions*, Holt, Rinehart and Winston, 1966.

[10] J. Lehner, *Discontinuous groups and automorphic functions*, Mathematical Surveys 8, American Math. Soc., 1964.

[11] W. Magnus, *Non-Euclidean tesselations and their groups*, Academic Press, 1974.

[12] B. Maskit, *Kleinian groups*, Springer, 1988.

The virtual cohomological dimension of Coxeter groups

Mladen Bestvina

Mathematics Department, UCLA, Los Angeles, California, CA 90024, USA.

Abstract. For every finitely generated Coxeter group Γ we construct an acyclic complex of dimension vcd Γ where Γ acts cocompactly as a reflection group with finite stabilizers. This provides an effective calculation of vcd Γ in terms of the Coxeter diagram of Γ.

Introduction

A *Coxeter system* is a pair (Γ, V) where Γ is a group (called a *Coxeter group*) and V is a finite set of generators for Γ all of which have order two, such that all relations in Γ are consequences of relations of the form $(vw)^{m(v,w)} = 1$ for $v, w \in V$ and $m(v, w)$ denotes the order of vw in Γ. In particular, $m(v, v) = 1$ and $m(v, w) = m(w, v) \in \{2, 3, \ldots, \infty\}$.

In this note we address the question (see [Pr, problem 1]): What is the virtual cohomological dimension (vcd) of a Coxeter group?

Every Coxeter group can be realized as a group of matrices (see [Bou]), and consequently has a torsion free subgroup of finite index (by Selberg's lemma). Davis [Dav] has constructed a finite dimensional contractible complex (we review the construction below) where a given Coxeter group acts properly discontinuously, so that any torsion free subgroup acts freely. It follows from these remarks that the vcd of any Coxeter group is finite.

All finite Coxeter groups have been classified (see [Bou]), thus providing the answer to the above question when vcd equals zero. Another special case (when vcd ≤ 1) was resolved in [Pr-St].

By $\mathcal{F}$ denote the set of all subsets of V (including the empty set) that generate a finite subgroup of Γ (recall that for any subset F of V and the subgroup

Supported in part by the Alfred P. Sloan Foundation, the Presidential Young Investigator Award, NSF and the IHES.

$\langle F \rangle$ of Γ generated by F, the pair $(\langle F \rangle, F)$ is a Coxeter system). Mike Davis has worked out a method in [Dav] for constructing polyhedra where Γ acts as a reflection group. One produces a *panel complex* $(K, \mathcal{P})$, i.e. a compact polyhedron K and a collection $\mathcal{P} = \{P_F \subseteq K \mid F \in \mathcal{F}\}$ of subpolyhedra (called *panels*) with the following properties:

1. $P_\varnothing = K$,
2. each P_F is acyclic (and in particular non-empty), and
3. $P_{F_1} \cap P_{F_2}$ equals $P_{F_1 \cup F_2}$ when $F_1 \cup F_2 \in \mathcal{F}$ and it is empty otherwise.

Then Γ acts as a reflection group, cocompactly and properly discontinuously on the acyclic polyhedron $X = \Gamma \times K/\sim$, where $(\gamma_1, x_1) \sim (\gamma_2, x_2) \Leftrightarrow x_1 = x_2$ and $\gamma_1^{-1}\gamma_2 \in \langle v \in V \mid x_1 = x_2 \in P_v \rangle$ (see [Dav, Theorem 10.1]).

Davis constructs a natural panel complex $(K_D, \mathcal{P}_D)$ associated with a Coxeter system (Γ, V) as follows. K_D is the (geometric realization of the) complex whose set of vertices is $\mathcal{F}$, and a subset of $\mathcal{F}$ spans a simplex if and only if it is linearly ordered by inclusion. K_D is a cone, with $\varnothing$ as the cone point. The panel P_F associated with $F \in \mathcal{F}$ is the union of the simplices of the form $\langle F_1 \subset F_2 \subset \cdots \subset F_k \rangle$ with $F \subseteq F_1$, and it is also a cone, with cone point F. The resulting polyhedron $X_D = \Gamma \times K_D/\sim$ is contractible, and lends itself to geometric investigation, due to the fact that it supports a metric of non-positive curvature (see [Mou]). However, for the purposes of calculating the virtual cohomological dimension of Γ, this polyhedron is inappropriate, as its dimension, which always provides an upper bound for vcd Γ (see [Dav, Proposition 14.1]), is too large. For example, when Γ is finite, vcd $\Gamma = 0$ while $\dim X_D = \operatorname{card} V$.

In this note we propose a different panel complex $(K, \mathcal{P})$ with $\dim K = \operatorname{vcd}\Gamma$. First let us give an alternative description of the Davis panel complex $(K_D, \mathcal{P}_D)$.

With respect to inclusion, $\mathcal{F}$ is a partially ordered set. For every maximal element $F \in \mathcal{F}$ define P_F to be a point. Let $F \in \mathcal{F}$ be any element. Assuming that $P_{F'}$ has been defined for every $F' \supset F$, define P_F to be the cone on $\bigcup_{F' \supset F} P_{F'}$. In particular, $K_D = P_\varnothing$ is the cone on $\bigcup_{F \neq \varnothing} P_F$.

We now describe a simple modification in the above procedure that yields a polyhedron of optimal dimension.

1. The construction

For every maximal element $F \in \mathcal{F}$ define P_F to be a point. Let $F \in \mathcal{F}$ be any element. Assuming that $P_{F'}$ has been defined for every $F' \supset F$, define P_F to be *an acyclic polyhedron containing* $\bigcup_{F' \supset F} P_{F'}$ *of the least possible dimension*. Of course, most of the time P_F is just the cone on $\bigcup_{F' \supset F} P_{F'}$, but there is one case when we can get away with $\dim P_F = \dim \bigcup_{F' \supset F} P_{F'}$.

Lemma. *If L is a compact n-dimensional polyhedron with $H_n(L; \mathbb{F}) = 0$ for every field $\mathbb{F} \in \{\mathbb{Q}\} \cup \{\mathbb{Z}_p \mid p \text{ prime}\}$, then L embeds in a compact acyclic n-dimensional polyhedron as a subpolyhedron.*

Proof. First note that it suffices to prove the lemma for $n > 1$ and for $(n-2)$-connected L. (Set $L_{-1} = L$ and inductively for $i = 0, 1, 2, \cdots, n-2$, let L_i be L_{i-1} with the cone on its i-skeleton attached. Simply replace L with L_{n-2}.) The universal coefficients theorem implies that $H_n(L; \mathbb{Z}) = 0$ and that $H_{n-1}(L; \mathbb{Z})$ is a free abelian group. If $n > 2$, $\pi_{n-1}(L) \cong H_{n-1}(L; \mathbb{Z})$, and we can attach n-cells to L, one for each basis element of $\pi_{n-1}(L)$, to obtain a contractible n-dimensional polyhedron containing L. For $n = 2$, the Hurewicz homomorphism $\pi_1(L) \to H_1(L; \mathbb{Z})$ is surjective, and we can attach 2-cells to L killing $H_1(L; \mathbb{Z})$ without changing $H_2(L; \mathbb{Z})$. The attaching maps can be chosen to be PL, so the resulting space is a polyhedron.

Continue the construction of the P_F's until all panels are built. The last panel is the one associated with $\varnothing$, and it is the desired polyhedron K.

Theorem. $\text{vcd}\,\Gamma = \dim K$.

Proof. Let $n = \dim K$. Since any torsion-free subgroup of Γ acts freely and properly discontinuously on the acyclic n-dimensional polyhedron $X = \Gamma \times K / \sim$, it follows that $\text{vcd}\,\Gamma \leq n$ (since the simplicial chain complex for X provides a free resolution of $\mathbb{Z}$ of length n). To prove that the equality holds, we produce a locally finite non-trivial cycle in $H_n^{lf}(X; \mathbb{F})$. We can use this free resolution to compute group cohomology of a torsion-free subgroup Γ' of finite index. We obtain $H^n(\Gamma', \mathbb{F}\Gamma') = H_c^n(X, \mathbb{F})$, which is nonzero since the pairing $H_c^n(X, \mathbb{F}) \otimes H_n^{lf}(X; \mathbb{F}) \longrightarrow \mathbb{F}$ is non-degenerate.

First, consider a special case: assume that the last step in the construction of K involved coning, so that $\dim\left(\bigcup_{\varnothing \neq F \in \mathcal{F}} P_F\right) = n - 1$. That means that $H_{n-1}\left(\bigcup_{\varnothing \neq F \in \mathcal{F}} P_F; \mathbb{F}\right) \neq 0$ for some field $\mathbb{F}$, or equivalently, that

$$H_n\left(K, \bigcup_{\varnothing \neq F \in \mathcal{F}} P_F; \mathbb{F}\right) \neq 0.$$

Let C be a non-trivial relative cycle. Then $\tilde{C} = \sum_{\gamma \in \Gamma} (\text{sgn}\,\gamma)\,\gamma(C)$, where sgn : $\Gamma \to \{\pm 1\}$ is the orientation homomorphism, is a locally finite non-trivial cycle in $H_n^{lf}(X; \mathbb{F})$. (It is a cycle since ∂C is a chain of simplices whose stabilizers are finite and half of the elements are orientation reversing. It is non-trivial since it evaluates non-zero on each cocycle in $H^n(K, \bigcup_{\varnothing \neq F \in \mathcal{F}} P_F; \mathbb{F}) \subseteq H_c^n(K; \mathbb{F})$ on which C evaluates non-trivially.)

Now, for the general case, suppose that in the construction of the panels, P_{F_0} is the first panel with $\dim P_{F_0} = n$. Therefore, P_{F_0} is the cone on $\bigcup_{F \supset F_0} P_F$ and $H_{n-1}(\bigcup_{F \supset F_0} P_F; \mathbb{F}) \neq 0$ for some field $\mathbb{F}$. Consider the subgroup $\Gamma' \subset \Gamma$ generated by $V' = (\bigcup\{F \in \mathcal{F} \mid F_0 \subset F\}) \smallsetminus F_0$. Again, let

C represent a non-trivial relative cycle in $H_n(P_{F_0}, \bigcup_{F \supset F_0} P_F; \mathbb{F})$ and define $\tilde{C} = \sum_{\gamma \in \Gamma'}(\operatorname{sgn}\gamma)\,\gamma(C)$. Then $\tilde{C}$ is a non-trivial cycle in $H_n^{lf}(K; \mathbb{F})$ and thus $\operatorname{vcd}\Gamma = n$.   □

2. Example

Let $V = \{a, b, c, x, y, z\}$ with the Coxeter diagram as in Fig. 1 (if two genera-tors are not connected by an edge, the order of their product is infinite, and if they are connected by an edge, the order of their product is 2). The maximal elements of $\mathcal{F}$ are $\{a, x\}, \{b, y\}, \{c, z\}, \{x, y, z\}$. The construction yields that $\operatorname{vcd}\Gamma = 1$, and $(K, \mathcal{P})$ is pictured in Fig. 2. The label of a vertex or an edge is the subset of V consisting of those $v \in V$ such that $P_{\{v\}}$ contains that vertex or edge.

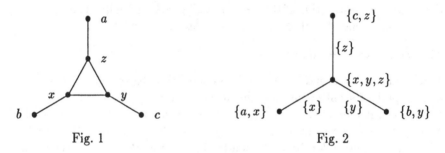

Fig. 1   Fig. 2

Remarks.

(1) The same procedure can be used to compute $\operatorname{vcd}_{\mathbb{F}} \Gamma$ for any field $\mathbb{F}$.

(2) When $\operatorname{vcd}\Gamma \neq 2$, the above construction produces a *contractible* polyhe-dron of dimension $\operatorname{vcd}\Gamma$ where Γ acts as a reflection group. I do not know whether such a polyhedron exists when $\operatorname{vcd}\Gamma = 2$. Concretely, let K be a non-simply-connected acyclic 2-dimensional full simplicial complex with the vertex set V. For $v \in V$ let P_v be the star of v with respect to the first barycentric subdivision of K, and let $m(v, w) = 2$ when v and w are con-nected by an edge, and $m(v, w) = \infty$ otherwise. As in [Dav] this induces a right-angled Coxeter group whose virtual cohomological dimension is 2. I do not know whether or not a torsion-free subgroup G of finite index admits a 2-dimensional $K(G, 1)$.

(3) The construction of (2) for a full triangulation K of the projective plane $\mathbb{R}P^2$ yields a Coxeter group Γ. It is not hard to show that $\operatorname{vcd}\Gamma = 3$ while $\operatorname{vcd}_{\mathbb{Q}}\Gamma = 2$. A torsion-free negatively curved group with the same properties was constructed in [B-M].

References

[B-M] M. Bestvina and G. Mess, *The boundary of negatively curved groups*, J. Amer. Math. Soc. (3) **4** (1991), pp. 469–481.

[Bou] N. Bourbaki, *Groupes et Algèbres de Lie*, Chapters IV-VI, Hermann, Paris, 1968.

[Dav] M. Davis, *Groups generated by reflections*, Annals of Math. **117** (1983), pp. 293–324.

[Mou] G. Moussong, *Hyperbolic Coxeter groups*, Ph.D. Thesis Ohio State University, 1988.

[Pr] S.J. Pride, *Groups – Korea 1988*, (A.C. Kim and B.H. Neumann, eds.), Lect. Notes in Mathematics 1398, Springer, Berlin.

[Pr-St] S.J. Pride and R. Stöhr, *The (co)homology of aspherical Coxeter groups*, preprint.

The Geometric Invariants of a Group
A Survey with Emphasis on the Homotopical Approach

Robert Bieri

Fachbereich Mathematik, Universität Frankfurt, Robert-Mayer-Str. 6, D 6000 Frankfurt 1, Germany.

1. Introduction

The idea behind the terminology "geometric invariant of a group" is borrowed from the founding fathers of homological algebra. We consider an abstract group G and let it act by covering transformations on the universal cover X of an Eilenberg-MacLane complex, $K(G, 1)$. Then we use X in order to define an object $\Sigma^*(G)$ which turns out to be independent of the particular choice of X, i.e. it is an invariant of G, and which contains rather detailed information on the internal structure of G. Finally, we take some care to find an algebraization of the construction $G \mapsto \Sigma^m(G)$ — not because we dislike the topological approach, but because we aim to understand it thoroughly by extending its scope from group rings to general rings and modules.

All this is modelled after the classical approach of Eckmann, Eilenberg-MacLane and Hopf to group cohomology. Our invariants $\Sigma^*(G)$, however, are not Abelian groups but form a descending chain of open subsets

$$\Sigma^0(G) \supseteq \Sigma^1(G) \supseteq \cdots \supseteq \Sigma^m(G) \supseteq \cdots$$

of a certain concrete space (in the most important case the $\mathbb{R}$-vector space $\mathrm{Hom}(G, \mathbb{R}^{add})$). This is the reason why we call our invariants "geometric". The issue becomes slightly confusing because they crop up in various brands according to their concrete definition: The homotopical ones ($\Sigma^*(G)$), the homological ones ($\Sigma^*(G; \mathbb{Z})$), and the algebraic ones ($\Sigma^*(G; A)$), where A stands for a G-module. But the situation is not too bad since the first two are closely related and the latter two actually coincide whenever both are defined.

The theory grew originally from joint effort with Ralph Strebel to understand the subtle difference between finitely presented and non-finitely presented metabelian groups in terms of their internal structure — the relevant information turns out to be contained in $\Sigma^1(G)$ or in $\Sigma^0(G/G';G')$. I owe Ralph a great amount of gratitude for the excellent collaboration over the past decade; and also to John Groves, Walter Neumann and Burkhardt Renz for their collaboration in the process of extending the theory to general groups and higher dimensions.

In this survey the emphasis is put on the homotopical approach due to Renz (see [R 87] and [R 88]). I feel that this is closest to the intuitive ideas and best to visualize. The generalization from translation action to arbitrary essentially isometric action, the concept of horo-connectivity and most of Theorem A is new; the proofs will appear elsewhere.

2. Essentially isometric action on $\mathbb{R}$

Let $g : \mathbb{R} \to \mathbb{R}$ be an orientation preserving homeomorphism of the real line. By the *defect* of g we mean the quantity

$$D_g := \sup\{|d(gI) - d(I)| \,\big|\, \text{all segments } I \subseteq \mathbb{R}\},$$

where $d(I)$ stands for the length of the segment I. Thus D_g measures to what extent g fails to be an isometry. Of course, D_g may be infinite, and if D_g is finite the set $\{D_{g^m} \mid m \in \mathbb{Z}\}$ may be unbounded.

Let G be a group acting on the real line, via a homomorphism $\rho : G \to \text{Homeo}^+(\mathbb{R})$. We say that this G-action ρ is *essentially isometric*, if there is a finite bound on $\{D_{\rho(g)} \mid g \in G\}$. We write $EI(G)$ for the set of all essentially isometric G-actions on $\mathbb{R}$. This is a subset of the set $\text{Hom}(G, \text{Homeo}^+(\mathbb{R}))$ of all G-actions by orientation preserving homeomorphisms and therefore inherits the compact-open topology (with respect to the discrete topology on G and the compact-open topology on the homeomorphism group of $\mathbb{R}$).

If one prefers to work with more concrete subspaces of $EI(G)$ one can fix a specific group H of essentially isometric homeomorphisms (with bounded defects!) and focus attention on $\text{Hom}(G, H)$. Interesting special cases are:

1. H is the translation group (i.e. $\cong \mathbb{R}^{add}$). If G is finitely generated then $\text{Hom}(G, \mathbb{R})$ becomes isomorphic to the real vector space $\mathbb{R}^n$, where n is the $\mathbb{Z}$-rank of the Abelianization G/G'.

2. $H = \widetilde{SL_2}(\mathbb{R})$, the universal cover of the Lie group $SL_2(\mathbb{R})$, regarded as a group of homeomorphisms on the circle S^1 of all directions in $\mathbb{R}^2$. (This has been suggested by G. Meigniez in [M]). The advantage over $\text{Hom}(G, \mathbb{R})$ is that commutator elements of G may now act non-trivially on $\mathbb{R}$.

3. H is any other group of homeomorphisms of $\mathbb{R}$ lifting a subgroup of $\text{Homeo}^+(S^1)$.

The action ρ of G on $\mathbb{R}$ is said to be *degenerate* if it satisfies one of the following three conditions (which are equivalent under the assumption that $\rho \in EI(G)$).

(a) Every G-orbit is bounded.

(b) Every element $g \in G$ has a fixed point.

(c) G has a global fixed point.

We will usually have to exclude degenerate actions. The set of all non-degenerate actions of $EI(G)$ is open in $EI(G)$.

3. The homotopical geometric invariant $\Sigma^*(G)$

Let G be a group. By an *Eilenberg-MacLane complex* for G, or a $K(G,1)$, we mean a connected CW-complex with fundamental group G and vanishing higher homotopy groups. Following C. T. C. Wall we say that G is *of type F_m*, $m \in \mathbb{N}_0$, if G admits an Eilenberg-MacLane complex Y with finite m-skeleton Y^m. Throughout the paper we will assume that the group G is of type F_m for some fixed m, and that X is the universal covering complex of a $K(G,1)$ with finite m-skeleton.

Let $\rho \in EI(G)$ be a non-degenerate essentially isometric action of G on $\mathbb{R}$. Any continuous equivariant map $h : X \to \mathbb{R}$ shall be called an *equivariant height function* on X. It is not difficult to see that equivariant height functions always exist and that any two such are *equivalent* on the m-skeleton X^m, in the sense that there is a constant $D \in \mathbb{R}$ such that $|h(x) - h'(x)| \le D$, for all $x \in X^m$. Moreover, by using a simplicial subdivision of the cells of X, one can actually find a continuous map from $\mathrm{Hom}(G, \mathrm{Homeo}^+(\mathbb{R}))$ into the space of all continuous functions $C(X, \mathbb{R})$ which picks, for each G-action ρ, a corresponding equivariant height function.

Given any G-equivariant height function $h : X \to \mathbb{R}$ we let $X_h \subseteq X$ be the maximal subcomplex of X contained in $h^{-1}([0, \infty))$.

One might be tempted to define a subset of $EI(G)$ by considering all non-degenerate essentially isometric G-actions on $\mathbb{R}$ with the property that X_h is k-connected. Unfortunately, the connectivity properties of X_h depend heavily on the choice of Y and h (it is only for $k = 0$ where one can make this work by restricting attention to the Cayley graph and take h to be linear on edges). Therefore we regretfully have to put

Definition. *The geometric invariant $\Sigma^m(G) \subseteq EI(G)$ is the set of all non-degenerate essentially isometric G-actions on $\mathbb{R}$ with the property that there is a suitable choice for $X = \tilde{Y}$ and $h : X \to \mathbb{R}$ such that X_h is $(m-1)$-connected.*

The connectivity condition is empty if $m = 0$ and so $\Sigma^0(G)$ consists of all non-degenerate actions in $EI(G)$. For completeness we put $\Sigma^m(G) = \varnothing$ if G is not of type F_m.

4. Criteria and openness result

As before, let ρ be a non-degenerate essentially isometric action of G on $\mathbb{R}$, let X be the universal cover of a $K(G,1)$ with finite m-skeleton and $h : X \to \mathbb{R}$ an equivariant height function. We shall now exhibit criteria for ρ to be in $\Sigma^m(G)$, in terms of connectivity properties which are weaker than $(m-1)$-connectivity of X_h but are independent of all choices. All these properties use, in a sense, the $\mathbb{R}$-graded filtration of X by $X_h^{[r,\infty)}$, the maximal subcomplex of X contained in $h^{-1}([r,\infty))$, for $r \in \mathbb{R}$.

A: Essential-connectivity. We say that $X_h^{[r,\infty)}$ is *essentially k-connected*, for some $k \geq -1$, if there is a real number $d \geq 0$ with the property that the map

$$\iota_* : \pi_i(X_h^{[r,\infty)}) \to \pi_i(X_h^{[r-d,\infty)}),$$

induced by inclusion, is zero (i.e. has singleton image) for all $i \leq k$. (The condition is empty for $i = -1$). It is not difficult to see that if $X_h^{[r,\infty)}$ is essentially k-connected for some $r \in \mathbb{R}$ then it is so for every $r \in \mathbb{R}$ with a uniform choice of d.

B: Horo-connectivity. Let Z be a locally compact topological space. We call a continuous map $f : Z \to X$ *horo-continuous* (with respect to h), if the subspaces $Z(r) = \{z \in Z \mid h(f(z)) \leq r\}$ are compact for all $r \in \mathbb{R}$.

Remark. Thus f is horo-continuous if it is proper with respect to two specific filtrations on Z and X, respectively: the filtration on Z given by an exhausting chain of compact neighbourhoods, and the filtration on X by $h^{-1}((-\infty,r])$, $r \in \mathbb{R}$, (or, equivalently, by the closure of the complements of $X_h^{[r,\infty)}$ in X). For the concept of proper maps and proper homotopy with respect to filtrations see the forthcoming book of Ross Geoghegan. Our terminology follows the intuition that sequences $\{x_i \in X \mid i \in \mathbb{N}\}$ with $\lim_{i \to \infty} h(x_i) = \infty$ "approach the boundary".

We now say that X is *horo-k-connected* (with respect to h), if each horo-continuous map $f : \mathbb{R}^i \to X$ extends to a horo-continuous map $\tilde{f} : \mathbb{R}^i \times [0,\infty) \to X$, for all $i \leq k$ and $k \geq 0$.

Theorem A. *Let G be a group of type F_m and $\rho : G \to \text{Homeo}^+(\mathbb{R})$ a non-degenerate essentially isometric action of G on $\mathbb{R}$. Let X be the universal cover of a $K(G,1)$ with finite m-skeleton and $h : X \to \mathbb{R}$ an equivariant height function. Then the following five conditions are equivalent*

(i) ρ is in $\Sigma^m(G)$.

(ii) X_h is essentially $(m-1)$-connected.

(iii) X is horo-m-connected.

(iv) There is a continuous cellular G-equivariant map $\varphi : X^m \to X^m$ with

$$h\big(\varphi(x)\big) \geq h(x) + \varepsilon \text{ for all } x \in X^m \text{ and some } \varepsilon > 0.$$

28 R. Bieri

(v) There are real numbers ε, μ with $\varepsilon > 0$, and a continuous cellular G-equivariant map $\sigma : X^m \times [0, \infty) \to X$ such that $\sigma(-, 0)$ is the embedding $X^m \subseteq X$ and

$$h\big(\sigma(x,t)\big) \geq h(x) + \varepsilon t + \mu, \quad \text{for all } x \in X^m, \ t \in [0, \infty).$$

Corollary A1. *The assertions $(i) - (v)$ in the theorem are independent of the particular choice of X and $h : X \to \mathbb{R}$.* □

For every real number $D \geq 0$ we consider the subset $EI_D(G) \subseteq EI(G)$ consisting of all G-actions with defects *globally bounded* by D. Note that, e.g. $\text{Hom}(G, \widetilde{SL_2}(\mathbb{R})) \subseteq EI_{2\pi}(G)$, so that the following corollary answers Meigniez' question 2 on p. 305 in [M].

Corollary A2. *For each $D \geq 0$ the subset $\Sigma^m(G) \cap EI_D(G)$ is open in $EI_D(G)$.*

Proof. Let $\rho \in EI(G)$ be in this subset. Upon replacing the map φ by a power φ^p, if necessary, we may make the number ε in condition (iv) as big as we wish; we want it to be $\varepsilon = D + 3\varepsilon'$ for some $\varepsilon' > 0$. Next we choose a compact subset $C \subseteq X^m$ with $GC = X^m$ and a continuous map $EI(G) \to C(X, \mathbb{R})$ which assigns to each $\rho' \in EI(G)$ an equivariant height function $h' : X \to \mathbb{R}$. We can then choose ρ' sufficiently close to ρ so that the difference of h and h', when restricted to C, is at most ε'. This implies that

$$|h'(\varphi(c)) - h'(c)| \geq D + \varepsilon', \quad \text{for all } c \in C.$$

As D is a global bound for the defects it follows that

$$|h'(\varphi(x)) - h'(x)| \geq \varepsilon', \quad \text{for all } x \in X.$$

Hence (iv) still holds true for ρ replaced by ρ' and ε replaced by ε'. This shows that ρ' is in $\Sigma^m(G)$, and proves the assertion. □

5. The homological geometric invariant $\Sigma^*(G; \mathbb{Z})$

It is a straightforward matter to write down the homological version of the homotopical concepts of Sections 3 and 4. As above, let X denote the universal cover of a $K(G, 1)$ with finite m-skeleton and $h : X \to \mathbb{R}$ a G-equivariant height function. Along the lines of Section 3 we define $\Sigma^m(G; \mathbb{Z})$ to be the set of all non-degenerate essentially isometric G-actions on $\mathbb{R}$ with the property that there is a suitable choice for X and $h : X \to \mathbb{R}$ such that X_h is $(m - 1)$-acyclic. (The notation involving $\mathbb{Z}$ will turn out to be convenient later.)

From this definition it is obvious that one has an inclusion $\Sigma^m(G) \subseteq \Sigma^m(G; \mathbb{Z})$, for each m, since $(m-1)$-connected spaces are $(m-1)$-acyclic. Moreover, this inclusion is an equality for $m = 0$ and $m = 1$. That is trivial for $m = 0$ and obvious for $m = 1$ since the concepts of connectivity and 1-acyclicity coincide for CW-complexes. It is conceivable that $\Sigma^m(G) = \Sigma^m(G; \mathbb{Z})$ for all m, but that is an open problem. To prove this for $m = 2$ would imply it for arbitrary m because we can show

Theorem B. *If G is a group of type F_m then $\Sigma^m(G) = \Sigma^2(G) \cap \Sigma^m(G; \mathbb{Z})$.*

Sketch of proof. Assume $\rho \in \Sigma^2(G)$. By definition one can choose X and h such that X_h is 1-connected. If ρ is also in $\Sigma^m(G; \mathbb{Z})$, then X_h is essentially $(m-1)$-acyclic by Theorem A. By the method which is also used to prove the implication (ii) $\Rightarrow$ (i) in Theorem A, one can modify X, in terms of elementary expansions in the sense of simple homotopy theory, so that X_h remains 1-connected but is also $(m-1)$-acyclic. The Hurewicz Theorem then asserts that X_h is $(m-1)$-connected, i.e. $\rho \in \Sigma^m(G)$. This establishes the non-trivial inclusion of Theorem B. $\square$

The proof of Theorem A carries over to the homological invariant $\Sigma^*(G; \mathbb{Z})$. We may omit the proof at this stage since it will be covered by the algebraic result Theorem C. But it is most instructive to translate, at least intuitively, the concepts of essential connectivity and horo-connectivity. Clearly X_h should be called *essentially k-acyclic* if there is a real number $d \geq 0$ such that inclusion induces the zero map

$$\tilde{H}_i(X_h) \to \tilde{H}_i(X_h^{[-d,\infty)}),$$

for all $i \leq k$ (reduced homology). To translate horo-connectedness into horo-acyclicity is more interesting: Corresponding to continuous maps $f : \mathbb{R}^i \to X$ we consider infinite locally finite i-chains of X (ordinary i-chains would correspond to continuous maps $S^i \to X$). Let $\mathbf{C}(X) \subseteq \mathbf{C}^\infty(X)$ denote the cellular chain complexes of ordinary and infinite chains, respectively. Every chain $c \in \mathbf{C}^\infty(X)$ is carried by its *support* $\mathrm{supp}(c) \subseteq X$ which is the union of all cells involved in c. Now, the condition on a chain $c \in \mathbf{C}^\infty(X)$ which would correspond to horo-continuity of a map $f : \mathbb{R}^i \to X$ is, that the intersections $\mathrm{supp}(c) \cap h^{-1}((-\infty, r])$ are compact for all $r \in \mathbb{R}$. In other words, c involves, for every $r \in \mathbb{R}$, only a finite number of cells e outside $X_h^{[r,\infty)}$. Let us write $\hat{\mathbf{C}}(X)$ for the subcomplex of all chains c with this property. Then X should and will be termed *horo-k-acyclic with respect to h* if $H_i(\hat{\mathbf{C}}(X)) = 0$, for all $i \leq k$.

In order to prepare algebraization of these concepts we have to replace all data of the space X by data on the chains of X. In particular, we can replace the equivariant height function $h : X \to \mathbb{R}$ by the map $v : \mathbf{C}(X) \to \mathbb{R}_\infty$ given by $v(c) = \inf h(\mathrm{supp}(c))$. Here, $\mathbb{R}_\infty$ stands for the real numbers supplemented by an element ∞ which satisfies $r \leq \infty$ and $r + \infty = \infty$ for every $r \in \mathbb{R}_\infty$; and inf $\varnothing = \infty$. One observes readily that both essential- and horo-acyclicity can be rephrased in terms of v.

6. Algebraization

We are now in a position to introduce the algebraical geometric invariants $\Sigma^*(G; A)$. We do this in two steps. First we mimic the situation of the cellular chain complex $\mathbf{C}(X)$ of Section 5 in an arbitrary $\mathbb{Z}G$-free resolution; then we reinterpret the result in more familiar algebraic terms.

6.1. As before, let G act on $\mathbb{R}$ by some non-degenerate action $\rho \in EI(G)$. Let A by a G-module. By a *valuation on A* we mean an equivariant map $v : A \to \mathbb{R}_\infty$ satisfying

(i) $v(a + b) \geq \inf \{v(a), v(b)\}$,
(ii) $v(-a) = v(a)$,
(iii) $v(0) = \infty$.

for all $a, b \in A$. If v and w are two valuations on A, we say that v *is dominated by w*, (written $v \leq w$), if there is some $r \in \mathbb{R}$ such that $v(a) \leq w(a) + r$, for all $a \in A$. If v and w dominate each other we say that they are *equivalent*.

Valuations can rather easily be constructed on a free G-module F. Indeed, if $\mathcal{X}$ is a basis of F then every map $v : \mathcal{X} \to \mathbb{R}$ can first be extended to a G-map $v : G\mathcal{X} \to \mathbb{R}$ and then, by putting

$$v(c) = \inf v(\mathrm{supp}(c)), \qquad c \in F,$$

to a valuation on F; here $\mathrm{supp}(c)$ stands for the set of all elements of the $\mathbb{Z}$-basis $G\mathcal{X}$ occurring with non-zero coefficient in the expansion of c. Let us call a valuation which is obtained in this way *naive* (with respect to $\mathcal{X}$). One observes readily that if $\mathcal{X}$ is finite then v is dominated by every other valuation on F. This applies, in particular, to every naive valuation with respect to some other basis $\mathcal{X}'$. Hence the naive valuations on a finitely generated free G-module F define a canonical equivalence class. Associated to this we thus have the unique equivalence class of filtrations given by

(6.1) $F_v^{[r,\infty)} := \{c \in F \mid v(c) \geq r\}, \qquad r \in \mathbb{R}.$

A G-module A is said to be of type FP_m if it admits a free resolution $\mathbf{F}$

$$\cdots \to F_i \overset{\partial}{\to} F_{i-1} \to \cdots \to F_0 \to A \to 0,$$

with finitely generated m-skeleton $\mathbf{F}^{(m)} := \bigoplus_{i=0}^{m} F_i$. In order to mimic the topological situation we may assume that $\mathbf{F}$ is given with a $\mathbb{Z}G$-basis $\mathcal{X} \subseteq \mathbf{F}$

with the property that $0 \notin \partial \mathcal{X}$. It is clear that such resolutions always exist and we call them *admissible*. As naive valuations can achieve arbitrary prescribed values on $\mathcal{X}$ we can pick naive valuations $v_i : F_i \to \mathbb{R}_\infty$ in such a way that $v_{i-1}\partial \geq v_i$, for all i. Let $v : \mathbf{F} \to \mathbb{R}_\infty$ denote the collection of all v_i. Note that each $\mathbf{F}_v^{[r,\infty)}$ is a sub-chain complex of $\mathbf{F}$; for convenience we abbreviate $\mathbf{F}_v := \mathbf{F}_v^{[0,\infty)}$. Furthermore, we write $\tilde{\mathbf{F}}$ or $\tilde{\mathbf{F}}_v$ for the corresponding augmented chain complexes with A in dimension -1, endowed with the zero valuation $v_{-1}(A - \{0\}) = 0$.

The topological concepts of Sections 3–5 are now easily translated. We start with the *definition* of $\Sigma^m(G; A)$ by putting $\rho \in \Sigma^m(G; A)$ if and only if there is a suitable choice for $\mathbf{F}$ and v, as above, such that $\tilde{\mathbf{F}}_v$ is exact in all dimensions less than m.

Next, we say that $\mathbf{F}_v^{[r,\infty)}$ is *essentially k-acyclic*, for $k \geq -1$, if there is a real number $d \geq 0$ such that every i-cycle z of $\tilde{\mathbf{F}}$, with $-1 \leq i \leq k$, is the boundary of an $(i+1)$-chain c of $\tilde{\mathbf{F}}$ with $v(c) \geq v(z) - d$.

Finally, we consider the set $\hat{\mathbf{F}}$ of all formal sums $\sum_{i=0}^\infty n_i b_i$ with $n_i \in \mathbb{Z}$ and $b_i \in G\mathcal{X}$, such that $\{i \mid n_i \neq 0,\ v(b_i) \leq r\}$ is finite for each $r \in \mathbb{R}$. Then $\hat{\mathbf{F}}^{(m)}$ is a chain complex containing $\mathbf{F}^{(m)}$ as a subcomplex and at the expense of restricting attention to locally finite chains one can extend this to the $(m+1)$-skeleton. We say that $\mathbf{F}$ is *horo-k-acyclic*, for some k with $0 \leq k \leq m$, if $\hat{\mathbf{F}}$ is exact in all dimensions less than or equal k (including zero!).

Exactly parallel to Theorem A one can now state and prove

Theorem C. *Let A be a G-module of type FP_m and $\mathbf{F} \twoheadrightarrow A$ an (admissible) free resolution with finitely generated m-skeleton $\mathbf{F}^{(m)}$. If $\rho \in EI(G)$ is non-degenerate and $v : \mathbf{F} \to \mathbb{R}_\infty$ is a naive valuation on $\mathbf{F}$, then the following conditions are equivalent:*

(i) $\rho \in \Sigma^m(G; A)$,

(ii) $\tilde{\mathbf{F}}_v$ is essentially $(m-1)$-acyclic,

(iii) $\mathbf{F}$ is horo-m-acyclic,

(iv) there is a G-equivariant chain map $\varphi : \mathbf{F} \to \mathbf{F}$ lifting the identity of A with $v(\varphi(c)) \geq v(c) + \varepsilon$ for all $c \in \mathbf{F}^{(m)}$ and some $\varepsilon > 0$.

The equivalence of (iii) and (iv) for the case of translation action is due to J. C. Sikorav [S]. The remaining assertion (v) of Theorem A seems rather less interesting in the homological context. As in Section 4 we deduce

Corollary C1. *The assertions (i)–(iv) in Theorem C are independent of the particular choice of $\mathbf{F}$ and v.*

Corollary C2. $\Sigma^m(G; A) \cap EI_D(G)$ *is open in* $EI_D(G)$.

Moreover, if G is of type F_m and A is the trivial G-module $\mathbb{Z}$ then, choosing $\mathbf{F} = \mathbf{C}(X)$ (as in Section 5) and using either of (ii), (iii) or (iv) to describe

$\Sigma^m(G; \mathbb{Z})$, we can show that this algebraically defined invariant coincides with the homological one defined in Section 5.

6.2. So far we have been following as closely as possible the guideline provided by the topological arguments. It is now time to give up this principle and reinterprete the results in more traditional algebraic terms.

In the group G we consider the submonoid $G_\rho = \{g \in G \mid gr \geq r \text{ for all } r \in \mathbb{R}\}$. First, we observe that (6.1) is actually a filtration in terms of G_ρ-submodules. Next, we pick any element $g \in G_\rho$ which has no fixed points in $\mathbb{R}$ and obtain an ascending chain $G_\rho \subseteq G_\rho g^{-1} \subseteq G_\rho g^{-2} \subseteq \ldots$ exhausting G. This shows that the group ring $\mathbb{Z}G$ is an ascending union of cyclic free $\mathbb{Z}G_\rho$-modules and hence $\mathbb{Z}G$ is flat over $\mathbb{Z}G_\rho$. We are now in a position to prove

Proposition D1. *$\rho \in \Sigma^m(G; A)$ if and only if the G-module A is of type FP_m over $\mathbb{Z}G_\rho$.*

Proof. We have verified the assumptions needed to apply K. S. Brown's method to deduce, from the essential acyclicity condition (ii) of Theorem C, that if $\rho \in \Sigma^m(G; A)$ then A is of type FP_m over $\mathbb{Z}G_\rho$ (see [BR], Appendix to 3). Conversely, assume that the G-module A admits a $\mathbb{Z}G_\rho$-free resolution $\mathbf{E} \twoheadrightarrow A$ which is finitely generated in dimensions less than or equal m. The tensor product with $\mathbb{Z}G$ over $\mathbb{Z}G_\rho$ then yields a $\mathbb{Z}G$-free resolution $\mathbf{F} \twoheadrightarrow \mathbb{Z}G \otimes_{G_\rho} A$ with finitely generated m-skeleton $\mathbf{F}^m$. But $\mathbb{Z}G \otimes_{G_\rho} A$ is easily seen to be isomorphic to A, since A is a G-module. Now using the naive valuation on $\mathbf{F}$, which is zero on the original basis of $\mathbf{E}$, yields $\mathbf{F}_v = \mathbf{E}$, whence $\rho \in \Sigma^m(G; A)$. $\square$

Let us now consider the filtration of the group G in terms of the subsets

$$G(n) := \{g \in G \mid g \cdot 0 \geq n\}, \qquad n \in \mathbb{N},$$

and let $\widehat{\mathbb{Z}G}$ denote the completion of the group ring with respect to the corresponding filtration of $\mathbb{Z}G$ by the $\mathbb{Z}G_\rho$-ideals $\mathbb{Z}G(n)$, $n \in \mathbb{N}$. $\widehat{\mathbb{Z}G}$ is a ring which contains $\mathbb{Z}G$ as a subring. It has an explicit description as the set of all formal series $\lambda = \Sigma n_g g$ over $\mathbb{Z}$ (i.e. functions $\lambda : G \to \mathbb{Z}$) with the property that each $G(n)$ contains almost all of the support $\text{supp}(\lambda)$. As usual one also obtains corresponding completions for (finitely presented) G-modules by tensoring with $\widehat{\mathbb{Z}G}$ over $\mathbb{Z}G$. In any case one can show that the complex $\mathbf{\hat{F}}$ of infinite chains used to define horo-acyclicity is essentially isomorphic to $\widehat{\mathbb{Z}G} \otimes_G \mathbf{F}$, whence

Proposition D2. *$\rho \in \Sigma^m(G; A)$ if and only if $\text{Tor}_i^{\mathbb{Z}G}(\widehat{\mathbb{Z}G}, A) = 0$ for all $0 \leq i \leq m$.*

In the case of translation actions this was proved by J.C. Sikorav [S].

7. Some applications

This section contains some applications which I find particularly attractive. All of them are concerned with the *translation subspace* of $EI(G)$, i.e. the set $\mathrm{Hom}(G, \mathbb{R})$ of all homomorphisms of G into the translation group $\mathbb{R}^{add}$ of $\mathbb{R}$. Let us throughout assume that G is finitely generated, so that $\mathrm{Hom}(G, \mathbb{R}) \cong \mathbb{R}^n$, where n is the $\mathbb{Z}$-rank of the Abelianization of G. More details and more applications can be found in the forthcoming set of notes [BS 92].

7.1. Normal subgroups. Let $N \triangleleft G$ be a normal subgroup with Abelian factor group $Q = G/N$, and let k be the $\mathbb{Z}$-rank of Q. Then the map induced by the projection $G \twoheadrightarrow Q$ embeds $\mathrm{Hom}(Q, \mathbb{R}) \cong \mathbb{R}^k$ into the translation subspace $\mathrm{Hom}(G, \mathbb{R}) \subseteq EI(G)$.

Theorem E. [R 88, BR] *Assume that the group G is of type F_m (resp. FP_m). Then the normal subgroup N is of type F_m (resp. FP_m) if and only if $\mathrm{Hom}(Q, \mathbb{R}) \subseteq \Sigma^m(G)$ (resp. $\subseteq \Sigma^m(G; \mathbb{Z})$).*

A similar result holds also for $\Sigma^m(G; A)$. The geometric invariants thus contain full information on the F_m-type (resp. FP_m) of normal subgroups above the commutator subgroup. The conjunction of Theorem E with Corollary A2 yields an openness result on the subset of those which are of type F_m (resp. FP_m).

Holger Meinert has applied Theorem E to compute the F_m-type of N in the case when G is the direct product of finitely many free groups [Mt]. It turns out that FP_m coincides with F_m in this situation.

7.2. Finite presentations. One of the original motivations for introducing geometric invariants was the aim of understanding the impact of finiteness assumptions on the internal structure of a group. The statement of the following result uses the map $EI(G) \to EI(G)$, $\rho \mapsto -\rho$, where $-\rho : G \to \mathrm{Homeo}^+(\mathbb{R})$ is given by $(-\rho)(g)(r) = -\rho(g)(-r)$.

Theorem F. [BS 80] *Let G be a finitely generated group which contains no free subgroups of rank 2. If G is finitely presented then*

$$(7.1) \qquad \mathrm{Hom}(G, \mathbb{R}) \subseteq -\Sigma^1(G) \cup \Sigma^1(G).$$

Moreover, if G is metabelian then G is finitely presented if and only if (7.1) holds. □

The inclusion (7.1) is inherited by all factor groups of G, whence the

Corollary F1. *If a finitely presented group contains no free subgroups of rank 2 then its metabelian factor groups are finitely presented.*

The corollary is easier to appreciate for those who know Herbert Abels' example of a finitely presented soluble group with a non-finitely presentable centre-by-metabelian factor group [A 79]. Another instance where a converse of Theorem F is available is given by Abels' impressive result

Theorem F2. [A 87] *Let G be an S-arithmetic subgroup of a soluble connected algebraic group over a number field. Then G is finitely presented if and only if (7.1) holds and $H_2(G'; \mathbb{Z})$ is a finitely generated G/G'-module.*

It is conceivable that, for *metabelian groups* G, the invariant $\Sigma^1(G)$ contains the information whether G is of type F_m (resp. FP_m). We have reasons to think that the following might be the answer.

Conjecture. [B 81] *A metabelian group G is of type F_m if and only if the complement of $\Sigma^1(G)$ in $\mathrm{Hom}(G, \mathbb{R})$ has the property that every subset of m non-zero points is contained in an open half space of $\mathrm{Hom}(G, \mathbb{R})$.*

Note that Theorem F establishes the conjecture for $m = 2$. The conjecture has been proved to hold true for metabelian groups of finite rank (Åberg[Å]). For related Theorems "over a field" see [BG 82].

8. Polyhedrality

In view of the applications in Section 7 it is of considerable interest to compute the intersection of $\Sigma^1(G) \subseteq EI(G)$ with the translation subspace $\mathrm{Hom}(G; \mathbb{R}) \cong \mathbb{R}^n$ in specific situations. This seems not to be an easy matter in general. In each instance where we were able to do it, it turned out to be *polyhedral* (i.e. a finite union of finite intersections of open half spaces of $\mathbb{R}^n$). In most cases these polyhedral sets were *rationally defined* (i.e. the corresponding half spaces given by Diophantine inequalities); but non-rationally defined ones do exist (cf. [BNS], [BS 92]).

We mention two specific results; many more examples are to be found in [BS 92].

Theorem G1. [BG 84] *If G is a finitely generated metabelian group then $\Sigma^1(G) \cap \mathrm{Hom}(G, \mathbb{R})$ is polyhedral and rationally defined.*

The core of the proof is a curious result on Krull valuations on a field K: the set of all values that valuations $K \to \mathbb{R}_\infty$ can achieve on a fixed sequence $(a_1, a_2, \ldots, a_n) \in K^n$ is a rationally defined closed polyhedral subset of $\mathbb{R}^n$. If G is the fundamental group of a 3-manifold M then $\mathrm{Hom}(G; \mathbb{R}) \cong H^1(M; \mathbb{R})$ and its rational points, $\mathrm{Hom}(G, \mathbb{Z}) \cong H^1(M, \mathbb{Z})$, can be interpreted as the set of homotopy classes $[M, S^1]$. Those which correspond to fibrations are collected by certain faces of the unit sphere of Thurston's norm [T]. Based on this and Tischler [Ti], Walter Neumann proved

Theorem G2. [BNS] *If G is a 3-manifold group then* $\Sigma^1(G) \cap \mathrm{Hom}(G, \mathbb{R})$
*is the union of the open cones defined by all faces of the Thurston norm whose
rational points correspond to fibrations. Consequently, it is a disjoint union of
finitely many rationally defined convex polyhedral open cones and is invariant
under* $\rho \mapsto -\rho$.

Thurston has proved that any open subset of $\mathbb{R}^n$, as described in Theorem G2,
is the set $\Sigma^1(G) \cap \mathbb{R}^n$ for some 3-manifold group [T]. Similarly, one can show
that every open rationally defined polyhedral subset of $\mathbb{R}^n$ is the set $\Sigma^1(G) \cap$
$\mathbb{R}^n$ for some other group G (see [BNS], [BS 92]).

References

[A 79]   H. Abels, *An example of a finitely presented soluble group*, in: Homological Group
         Theory (C.T.C. Wall, ed.), London Math. Soc. Lecture Notes 36, Cambridge Uni-
         versity Press, 1979, pp. 205–211.

[A 87]   H. Abels, *Finite presentability of S-arithmetic groups — Compact presentability
         of solvable groups*, Lecture Notes in Mathematics 1261, Springer, 1987.

[Å]   H. Åberg, *Bieri-Strebel valuations (of finite rank)*, Proc. London Math. Soc. (3)
         52 (1986), pp. 269–304.

[B 81]   R. Bieri, *Homological dimension of discrete groups*, Queen Mary College Mathe-
         matics Notes, London, 1976, (2nd ed. 1981).

[BG 82] R. Bieri and J.R.J. Groves, *Metabelian groups of type* $(FP)_\infty$ *are virtually of type*
         (FP), Proc. London Math. Soc. (3) **45** (1982), pp. 365–384.

[BG 84] R. Bieri and J.R.J. Groves, *The geometry of the set of characters induced by
         valuations*, Journal für die reine und angew. Math. **347** (1984), pp. 168–195.

[BR]   R. Bieri and B. Renz, *Valuations on free resolutions and higher geometric invari-
         ants of groups*, Comment. Math. Helvetici **63** (1988), pp. 464–497.

[BS 80] R. Bieri and R. Strebel, *Valuations and finitely presented metabelian groups*, Proc.
         London Math. Soc. (3) **41** (1980), pp. 439–464.

[BS 81] R. Bieri and R. Strebel, *A geometric invariant for modules over an Abelian group*,
         Journal für die reine und angew. Math. **322** (1981), pp. 170–189.

[BS 82] R. Bieri and R. Strebel, *A geometric invariant for nilpotent-by-Abelian-by-finite
         groups*, Journal of Pure and Appl. Algebra 25 (1982), pp. 1–20.

[BS 92] R. Bieri and R. Strebel, *Geometric invariants for discrete groups*, to appear.

[BNS]   R. Bieri, W.D. Neumann and R. Strebel, *A geometric invariant of discrete groups*,
         Invent. Math. 90 (1987), pp. 451–477.

[BR 87] K.S. Brown, *Trees, valuations and the Bieri-Neumann-Strebel invariant*, Invent.
         Math. 90 (1987), pp. 479–504.

[M]   G. Meigniez, *Action de groupes sur la droite et feuilletage de codimension 1*, Thèse,
         Lyon (1988), and *Bouts d'un group opérant sur la droite I: Théorie algébrique*,
         Ann. Inst. Fourier 40, Grenoble, 1990, pp. 271–312.

36   R. Bieri

[Mt] H. Meinert, *The geometric invariants of direct products of virtually free groups*, preprint, Universität Frankfurt, 1990.

[R 87] B. Renz, *Geometric invariants and HNN-extensions*, in: Group Theory, Proceedings of Singapore conference (1987), de Gruyter, Berlin, 1989, pp. 465–484.

[R 88] B. Renz, *Geometrische Invarianten und Endlichkeitseigenschaften von Gruppen*, Dissertation, Frankfurt, 1988.

[S] J.-C. Sikorav, *Homologie de Novikov associée à une classe de cohomologie réelle de degré un*, preprint, 1989.

[St 84] R. Strebel, *Finitely presented soluble groups*, in: Group Theory — Essays for Philip Hall (K. W. Gruenberg, J. E. Roseblade, eds.), Academic Press, 1984, pp. 257–314.

[T] W.P. Thurston, *A norm on the homology of 3-manifolds*, Memoirs of the AMS 339, American Math. Soc., 1986, pp. 99–130.

[Ti] D. Tischler, *On fibering certain foliated manifolds over S^1*, Topology 9 (1970), pp. 153–154.

String Rewriting – A Survey for Group Theorists

Daniel E. Cohen

Department of Mathematics, Queen Mary and Westfield College, Mile End Road, London, E1 4NS.

Term rewriting can be described as the theory of normal forms in algebraic systems. The related notion of *string rewriting* is the theory of normal forms in presentations of monoids. Both these theories, especially the latter, should therefore be of interest to group theorists. However, they are better known to computer scientists, who use them to discuss such matters as automatic theorem proving.

With the present employment situation, as jobs are easier to find in computer science than in mathematics, young mathematicians may be glad to find branches of computer science with an algebraic flavour, and rewriting systems are of this nature. One of the experts in the field, Ursula Martin, began her mathematical career as a group theorist, and others who started in group theory have worked with her.

Recall that a *monoid* is a set with an associative multiplication and an identity (thus monoids differ from groups because elements need not have inverses). For any set X, the *free monoid X^** on X is the set of all finite sequences of elements of X (including the empty sequence) with the obvious multiplication. The empty sequence is the identity for this multiplication, so we usually denote it by 1. The elements of X^* are called *strings* or *words*.

A *rewriting system* $\mathcal{R}$ on X is a subset of $X^* \times X^*$. If $(l, r) \in \mathcal{R}$ then, for any strings u and v, we say that the string ulv *rewrites to* the string urv, and write $ulv \to urv$. For any string w, we say that w is *reducible* if there is a string z such that $w \to z$; if there is no such z we call w *irreducible*. We write $\overset{*}{\to}$ for the reflexive transitive closure of $\to$ and $\equiv$ for the equivalence relation generated by $\to$. We say that the strings u and v are *joinable*, written $u \downarrow v$, if there is a string w such that $u \overset{*}{\to} w$ and $v \overset{*}{\to} w$. We add the subscript $\mathcal{R}$ if it is necessary to look at two or more rewriting systems.

The quotient $X^*/\equiv$ is a monoid, which we call the monoid *presented by* $\langle X; \mathcal{R} \rangle$. The distinction between regarding $\mathcal{R}$ as giving a presentation and as being a rewriting system is that in the former case our attention is on $\equiv$ while in the latter it is on $\overset{*}{\to}$.

The example of a rewriting system most familiar in group theory comes in the definition of a free group. The free group on X can be regarded as the monoid $(X \cup X^{-1})^*/\equiv$, where $\equiv$ comes from the rewriting system $\{(xx^{-1}, 1), (x^{-1}x, 1);$ all $x \in X\}$. This example should be borne in mind when considering the results which follow.

If we are looking for normal forms for the elements of the monoid presented by $\langle X; \mathcal{R} \rangle$ the obvious choices are the irreducible elements. For this to be satisfactory, we would require that there is exactly one irreducible element in each equivalence class.

But there is an even more fundamental requirement; namely, that the normal form corresponding to an element can be obtained by repeated rewriting. Thus we call $\mathcal{R}$ *terminating* if there is no infinite sequence $w_1 \to w_2 \to \dots \to w_n \to \dots$ ($\mathcal{R}$ is also called *well-founded* or *noetherian*; we are using the notation in [14]). This is most easily achieved by requiring that $|l| > |r|$ for all $(l, r) \in \mathcal{R}$, in which case we call $\mathcal{R}$ *length-reducing*. It can also be achieved if $|l| \geq |r|$ for all $(l, r) \in \mathcal{R}$ and, further, if $|l| = |r|$ then r precedes l in the lexicographic order induced by some well-order on X; when this happens, we shall refer to $\mathcal{R}$ as a *lexicographic* rewriting system. The study of other sufficient conditions for termination in string rewriting and the more general term rewriting is a major research topic in the theory; see [13] for more on this.

A rewriting system is called a *Church-Rosser* system if $u \equiv v$ implies $u \downarrow v$, and is called *complete* if it is both terminating and Church-Rosser (such a system is sometimes called "canonical"; [14] suggests calling such a system "convergent", but I prefer the traditional word "complete"). Plainly an equivalence class in a Church-Rosser system contains at most one irreducible element. Also a terminating system in which each equivalence class contains only one irreducible element is Church-Rosser.

A rewriting system is called *confluent* if $w \overset{*}{\to} u$ and $w \overset{*}{\to} v$ implies $u \downarrow v$. It is easy to see that a system is Church-Rosser iff it is confluent. The neatest way of showing this is to observe that if the system is confluent then $\downarrow$ is transitive.

A rewriting system is called *locally confluent* if $w \to u$ and $w \to v$ implies $u \downarrow v$. A terminating locally confluent system is confluent. This result is sometimes known as the Diamond Lemma; it was first proved in [28] (see also [11] for a version of the proof, and [4] for further results). The proof is fairly easy, using an inductive principle which can be formulated for terminating systems.

Note that the results of the previous paragraphs apply to an arbitrary relation $\to$ on an arbitrary set, and do not require that $\to$ comes from a string rewriting system (or a term rewriting system). It is interesting to observe that this theory was used in [28] to obtain the Normal Form Theorem for free groups and also to obtain normal forms in the λ-calculus; the latter is closely related to the theory of functional programming in computer science.

Jantzen's book [20] contains numerous interesting results on string rewriting, and is a valuable reference. Complete rewriting systems for presenting various interesting groups are given in [25]. Small cancellation theory is also looked at in [25] and [3], where it is shown that the main results of that theory can be obtained in the current context. Jantzen's book contains a long list of references to results in the theory. Conferences on Rewriting Techniques and Applications are held regularly; their Proceedings are published in the Springer Lecture Notes in Computer Science, and usually include surveys as well as more technical material. Interesting papers have appeared in many computer science journals, such as *Journal of Symbolic Computation, Journal of Computer and System Sciences*, and *Information and Computation*.

There are many special results in the theory which are of interest to group theorists, and I mention only two of them. It is easy to see that the free abelian group of rank n cannot be generated as a monoid by n generators but it has a finite presentation on m generators if $m > n$. However [15], this group is presented by a finite complete rewriting system on m generators iff $m \geq 2n$.

An abelian subgroup of a free group is infinite cyclic, and an abelian subgroup of the free product of finite groups is either infinite cyclic or finite. Both these groups may be presented by a finite complete length-reducing rewriting system, namely $\{(xx^{-1}, 1), (x^{-1}x, 1) \mid x \in X\}$ for the free group on X, and $\{(ab, c) \mid a, b, c \in G_i$ for some i and $ab = c$ in $G_i\}$ on the alphabet $\bigcup G_i$ for the free product $* G_i$. Now let G be any group which can be presented by a finite complete length-reducing rewriting system. According to [26], any finitely generated abelian subgroup of G is either infinite cyclic or finite. If G has non-trivial centre then G itself is either infinite cyclic or finite. Also, if G has a non-trivial finite normal subgroup then G is finite.

It is not possible to decide whether or not a finite rewriting system is terminating, confluent, locally confluent, or complete (see [20] for details). However, if a finite system is known to be terminating we can decide whether or not it is complete, using a criterion due to Knuth and Bendix [23] given below.

Let $\mathcal{R}$ be a terminating system. For each string w choose an irreducible string $S(w)$ such that $w \xrightarrow{*} S(w)$. This can be done arbitrarily, but if we wish to perform an algorithmic process we could require $S(w)$ to be obtained by always rewriting the leftmost substring possible.

We call the triple of non-empty strings u, v, w an *overlap ambiguity* if there are r_1 and r_2 such that (uv, r_1) and (vw, r_2) are in $\mathcal{R}$; we then say that $r_1 w$ and $u r_2$ are the corresponding *critical pair*. The triple u, v, w of possibly empty strings is called an *inclusion ambiguity* if there are r_1 and r_2 (which must be distinct if both u and w are empty, but otherwise may be equal) such that (v, r_1) and (uvw, r_2) are in $\mathcal{R}$; we then say that ur_1w and r_2 are the corresponding critical pair. If p and q are the critical pair corresponding to an (overlap or inclusion) ambiguity u, v, w then $uvw \to p$ and $uvw \to q$.

Let $\mathcal{R}$ be a terminating system. For any critical pair p, q we have $S(p) \equiv S(q)$, and so, if $\mathcal{R}$ is complete then $S(p) = S(q)$. Conversely, if $S(p) = S(q)$ for all critical pairs p, q it is easy to see that $\mathcal{R}$ is locally confluent, and hence complete. Evidently, when $\mathcal{R}$ is finite we can decide whether or not this condition holds. Note that it would be enough to know that for any critical pair p, q we have $p \downarrow q$, which is sometimes easier to check.

The rewriting systems $\mathcal{R}$ and $\mathcal{S}$ are called *equivalent* if $\equiv_{\mathcal{R}}$ is the same as $\equiv_{\mathcal{S}}$; this condition is stronger than saying that the monoids presented by $\langle X; \mathcal{R} \rangle$ and $\langle X; \mathcal{S} \rangle$ are isomorphic. Any rewriting system is equivalent to a complete one, as is shown by the Knuth-Bendix procedure [23] given below.

Let $\mathcal{R}$ be a lexicographic system (and hence a terminating system). Let $\mathcal{R}'$ be obtained from $\mathcal{R}$ by considering all critical pairs p, q such that $S(p) \neq S(q)$ and adding to $\mathcal{R}$ for such a critical pair either $(S(p), S(q))$ or $(S(q), S(p))$; the choice of pairs to add is to be made so that $\mathcal{R}'$ remains lexicographic. Evidently $\mathcal{R}$ is complete iff $\mathcal{R}' = \mathcal{R}$. Because $S(p) \equiv_{\mathcal{R}} S(q)$, $\mathcal{R}'$ is equivalent to $\mathcal{R}$.

Now let $\mathcal{R}$ be an arbitrary system. Let $\mathcal{R}_0$ be obtained from $\mathcal{R}$ by replacing some of the pairs (l, r) by (r, l) in such a way that $\mathcal{R}_0$ is lexicographic. Inductively, define $\mathcal{R}_n$ for all n by $\mathcal{R}_{n+1} = (\mathcal{R}_n)'$, and let $\mathcal{R}_\infty = \bigcup_n \mathcal{R}_n$. It is easy to check that $\mathcal{R}_\infty$ is lexicographic and equivalent to $\mathcal{R}$. Consider a critical pair p, q for $\mathcal{R}_\infty$. Then p, q is a critical pair for $\mathcal{R}_n$ for some n. Letting $S_n(p)$ and $S_n(q)$ be the chosen irreducibles for $\mathcal{R}_n$ corresponding to p and q, we know that either $S_n(p) = S_n(q)$ or one of $(S_n(p), S_n(q))$ and $(S_n(q), S_n(p))$ is in $\mathcal{R}_{n+1}$. Hence $p \downarrow q$ for $\mathcal{R}_{n+1}$ and so also for $\mathcal{R}_\infty$. Thus, by the Knuth-Bendix criterion, $\mathcal{R}_\infty$ is complete.

The restriction to lexicographic systems is made to avoid problems with termination. If we simply know that $\mathcal{R}_n$ is terminating, we have to decide which of the two pairs to add in each case, and then try to show that $\mathcal{R}_{n+1}$ is also terminating. This process (in the more general case of term rewriting) has been much investigated, and there are computer programs aimed at performing this process either automatically or interactively, with backtracking as necessary (that is, if $\mathcal{R}_{n+1}$ is not terminating — or one cannot easily see that it is terminating — then we may reconsider the choices made for $\mathcal{R}_n$ or at earlier stages).

When $\mathcal{R}$ is finite then each $\mathcal{R}_n$ is finite, but $\mathcal{R}_\infty$ may be infinite. If, for some n, $\mathcal{R}_{n+1} = \mathcal{R}_n$ then $\mathcal{R}_n$ is complete and $\mathcal{R}_\infty = \mathcal{R}_n$. It follows that $\mathcal{R}$ is equivalent to the finite complete system $\mathcal{R}_n$. Conversely, if $\mathcal{R}$ is equivalent to some finite complete system then [25] there is some n such that $\mathcal{R}$ is equivalent to $\mathcal{R}_n$.

The rewriting system $\{(aba, bab)\}$ on the alphabet $\{a, b\}$ has no equivalent finite complete rewriting system [21]. However, an isomorphic monoid may be presented on the alphabet $\{a, b, c\}$ by the rewriting system $\{(ab, c), (ca, bc)\}$. This system is not complete, but the completion process gives the equivalent finite complete system $\{(ab, c), (ca, bc), (bcb, cc), (ccb, acc)\}$.

This example leads us to ask what properties are satisfied by a monoid which has (with respect to some set of generators) a presentation by a finite complete rewriting system. One necessary condition is that the monoid has a sovable word problem (recall that if the word problem is solvable in one finite presentation then it is solvable in every finite presentation). For let the monoid be presented by $\langle X; \mathcal{R} \rangle$, where $\mathcal{R}$ is a finite complete system. Since $\mathcal{R}$ is terminating and finite, we can calculate for each string u an irreducible string $S(u)$ such that $u \xrightarrow{*} S(u)$. Then the word problem for this presentation is solvable, since $u \equiv v$ iff $S(u) = S(v)$.

Groves and Smith [17] investigate how the property of being presented by a finite complete rewriting system behaves under various group-theoretic constructions (subgroups, quotient groups, wreath products, HNN extensions, etc.). In particular, they show that constructable soluble groups (see [2]) are presented by finite complete rewriting sytems; conversely, metabelian groups presented by finite complete rewriting systems are constructable, and they ask whether any soluble group presented by a finite complete rewriting system is constructable. They show that if A is a subgroup of the group G and A can be presented by a finite complete rewriting system then so can G if A has finite index in G and also if A is normal in G and such that G/A can be presented by a finite complete rewriting system. When A has finite index in G and G can be presented by a finite complete rewriting system it is not known whether A must be presentable by a finite complete rewriting system.

A break-through occurred in 1987 when Squier [29] showed that a monoid with a presentation by a finite complete rewriting system satisfies a homological condition, from which he was able to give examples of monoids with no such presentation. It was subsequently noticed that an earlier paper by Anick [1] had, using very different terminology, obtained a stronger homological condition.

A *free resolution* of a monoid M is an exact sequence

$$\ldots P_n \to P_{n-1} \to \ldots \to P_1 \to \mathbb{Z}M \to \mathbb{Z} \to 0$$

of free $\mathbb{Z}M$-modules. M is called FP_∞ if there is a free resolution such that P_i is finitely generated for all i, and it is called FP_n if there is a free resolution with P_i finitely generated for all $i \le n$. Squier proved that a monoid which can be presented by a finite complete rewriting system is FP_3, while Anick's result shows that such a monoid is FP_∞. For a monoid, as distinct from a group, we have to specify whether we are using left modules or right modules (an example of a monoid which is right FP_∞ but not left FP_1 is given in [10]), but since there is left-right symmetry in the definition of a finite complete rewriting system, these monoids are both left FP_∞ and right FP_∞.

Anick's result is much stronger, and obtains a free resolution from a (not necessarily finite) complete rewriting system. We say that a rewriting system is *reduced* if it has no inclusion ambiguities and for each $(l,r) \in \mathcal{R}$ the string r is irreducible. It is *strongly reduced* if it is reduced and each element of X is irreducible. It is easy [22] (se also [29]) to obtain a reduced complete rewriting system equivalent to a given complete rewriting system (if the original system is finite then the reduced system will also be finite), and we can immediately obtain a strongly reduced complete system which presents an isomorphic monoid (this system will not be equivalent to the previous one, since the set of generators is different).

Anick constructs a free resolution corresponding to a strongly reduced complete rewriting system. The free generators of each P_n are certain repeated overlaps, from which it is easy to see that the monoid is FP_∞ if the system is finite. Anick constructs the boundary maps and contracting homotopies in the resolution simultaneously by a complicated inductive process, involving not only boundary maps and contracting homotopies on elements of smaller degree but also their values on earlier elements of the same degree. The effect is that it is not possible to get a clear understanding of what is happening, and, although the construction can in principle be used for computation of homology, the definition of the boundary is too complicated to make his method practical. A subsequent generalisation and simplification by Koboyashi [24] still has the same problems.

The situation was elucidated by Brown [5] using a topological approach (the result was also proved by Groves [16], whose technique is intermediate between those of Anick and Brown). He showed that a strongly reduced complete rewriting system (in fact, he looked at the irreducibles rather than at the system itself) gives rise to a structure which he called a *collapsing scheme* on the bar resolution (which is a large resolution which can be obtained for any monoid), and that this collapsing scheme enables one to replace the bar resolution by a smaller resolution. The whole situation now becomes clear, and the boundary operators can be easily calculated. Thus the theorem not only enables us to use homological methods to obtain results about rewriting systems, but also enable us to use rewriting systems to prove results in the homology theory of groups. Brown's paper is very beautiful and contains some elegant calculations. In particular, the "big resolution" for the monoid $\langle x_i \ (i \in \mathbb{N}); \ (x_j x_i, x_i x_{j+1}) \text{ for } i < j \rangle$ constructed in [7] comes from the collapsing scheme corresponding to this complete rewriting system, and the proof that this monoid (and the group with the same presentation) is FP_∞ uses a new collapsing scheme on this resolution.

Brown's proof in [5] is topological, and he leaves it to the reader to translate the proof into an algebraic form. I think that there are likely to be people interested in Brown's work but unfamiliar with the topology. Such people should nonetheless have no difficulty in obtaining a general overview of the results from his account, and they should find the specific calculations straightforward and interesting. But I feel it is worthwhile to give an explicit algebraic translation of his proof, so that such readers do not have to take his main theorem on trust.

We begin with a chain complex

$$\mathbf{P} : \ldots \to P_{n+1} \overset{\partial_{n+1}}{\to} P_n \overset{\partial_n}{\to} P_{n-1} \to \ldots \to P_1 \to P_0$$

of free modules over a ring A. We require not only that each P_n is a free module but also that a specific basis is chosen for each P_n; the elements of this basis are called *n-cells*. A *collapsing scheme* on $\mathbf{P}$ is defined to consist of the following:

(1) a division of the cells into three pairwise disjoint classes, which we refer to as the *essential, redundant,* and *collapsible* cells, with all 0-cells being essential and all 1-cells being either essential or redundant;

(2) a function, called *weight*, from the set of all redundant cells into $\mathbb{N}$;

(3) a bijection, for each n, between the set of redundant n-cells and the set of collapsible $(n+1)$-cells, such that, if the collapsible cell c corresponds to the redundant cell r, then there is a unit u of A for which all redundant cells in the chain $r - u\partial c$ have weight less than the weight of r (in particular, if r has weight 0 then $r - u\partial c$ contains no redundant cells).

The motivation for this definition, and various examples, can be found in [5]. The weight is often given implictly, rather than explicitly, by the following procedure. Let S be an arbitrary set, and let $>$ be a binary relation on S such that, for all s, $\{t;\ s > t\}$ is finite (thus, we should think of $s > t$ as meaning, not "s is greater than t", but as "t is a child of s" or "t is used in s"). It is then a well-known (and easy to prove) result, sometimes called König's Lemma, that if, for some s, there are sequences $s > s_1 > \ldots > s_n$ for arbitrarily large n then there is an infinite sequence $s > s_1 > \ldots > s_n > s_{n+1} > \ldots$ It follows that if $>$ is terminating then, for each s there are only finitely many n for which there is a sequence $s > \ldots > s_n$, and the maximum such n may be called the weight of s. In particular, if $s > t$, then s has greater weight than t.

We can now state and prove the algebraic form of Brown's theorem. *If the free chain complex $\mathbf{P}$ has a collapsing scheme then it is chain-equivalent to a free chain complex $\mathbf{Q}$ for which the essential n-cells are a basis for Q_n for all n.* This holds for augmented complexes as well as for non-augmented ones.

We begin by defining homomorphisms $\theta_n : P_n \to P_n$ for all n as follows. For an essential cell e let $\theta e = e$ (subscripts will usually be omitted) and for a collapsible cell c let $\theta c = 0$. For a redundant cell r we define θr to be $r - u\partial c$, where c and u are as in (3) of the definition of a collapsing scheme (the definition would permit more than one suitable u; if this happens we just choose one). Let p be a chain in which all the redundant cells have weight at most k. Then θp is a chain in which all the redundant cells have weight less than k. We then see that all the redundant cells in $\theta^k p$ have weight 0, that $\theta^{k+1} p$ contains no redundant cells, and so $\theta^{k+2} p$ consists only of essential cells, and $\theta^m p = \theta^{k+2} p$ for $m > k + 2$. We can therefore define homomorphisms $\phi_n : P_n \to P_n$ by $\phi p = \theta^{k+2} p$ when all the redundant cells in p have weight at most k. Plainly $\phi = \phi\theta$.

We now show that $\phi\partial = \phi\partial\theta$. Since $\theta e = e$ we have $\phi\partial e = \phi\partial\theta e$. Since $\theta c = 0$, we need to show that $\phi\partial c = 0$; this holds because $\partial c = u^{-1}(r - \theta r)$ and $\phi = \phi\theta$. Finally, $r - \theta r = u\partial c$, so $\partial(r - \theta r) = 0$, and $\phi\partial r = \phi\partial\theta r$. It follows that $\phi\partial = \phi\partial\theta^m$ for all m, and hence $\phi\partial = \phi\partial\phi$.

Let Q_n be the free module with basis the essential n-cells. We can regard ϕ_n as a homomorphism from P_n to Q_n whenever convenient. We define $\delta_n : Q_n \to Q_{n-1}$ by $\delta = \phi\partial$. Then $\delta\delta = \phi\partial\phi\partial = \phi\partial\partial$, by the previous paragraph, so $\delta\delta = 0$, and we have a chain complex $\mathbf{Q}$. Also $\delta\phi = \phi\partial\phi = \phi\partial$, so ϕ is a chain-map from $\mathbf{P}$ to $\mathbf{Q}$. Notice that δ is easy to compute from ∂ and the collapsing scheme.

We next define homomorphisms $\alpha_n : P_n \to P_{n+1}$. We let $\alpha e = 0$ and $\alpha c = 0$ for any essential cell e or collapsible cell c. For a redundant cell r, we define αr using induction on the weight. Precisely, we have $r = u\partial c + \theta r$, where the redundant cells in θr have weight less than the weight of r, so we can define αr by $\alpha r = uc + \alpha\theta r$. Since αp involves only collapsible cells for any chain p, we have $\phi\alpha p = 0$ for all p.

Define $\psi_n : P_n \to P_n$ by $\psi = \iota - \alpha\partial - \partial\alpha$, where ι is the identity. We show that $\psi\theta = \psi$, from which it will follow, as before, that $\psi\phi = \psi$. First, $\psi\theta e = \psi e$, since $\theta e = e$. Next, $\alpha c = 0$, while $\partial c = u^{-1}(r - \theta r)$, so, by the definition of α, $\alpha\partial c = c$. Hence $\psi c = 0 = \psi\theta c$, since $\theta c = 0$. Finally, $\partial\alpha(r - \theta r) = r - \theta r$, by definition, and $\partial(r - \theta r) = \partial(u\partial c) = 0$, so $\psi(r - \theta r) = 0$. Also, plainly, $\psi\partial = \partial\psi$.

Since we may regard Q_n as a submodule of P_n, we may regard ψ_n as a homomorphism from Q_n to P_n. We then have $\partial\psi e = \psi\partial e = \psi\phi\partial e = \psi\delta e$, so that ψ is a chain-map from $\mathbf{Q}$ to $\mathbf{P}$.

Since $\alpha e = 0$, and $\phi\alpha p = 0$ for all p, we see that $\phi\psi e = e$. Also $\psi\phi = \psi = \iota - \alpha\partial - \partial\alpha$. Hence ϕ is a chain equivalence, as required.

Suppose that we have a strongly reduced complete rewriting system. We should compare the resolutions obtained by the various methods. It is easy to show that there is a natural bijection between the bases in Anick's resolution and Brown's. With more effort, using an inductive argument, we can show that the boundaries and contracting homotopies in Anick's resolution are the same as in Brown's (note that Brown's resolution has a contracting homotopy obtained from its chain-equivalence with the bar resolution and the natural contracting homotopy in the bar resolution). Because of the difficulty of explicitly constructing the boundaries in Anick's resolution, it is usually better to look at Brown's resolution when considering detailed questions of homology. However, I find Anick's description of the basis elements to be easier to follow than Brown's, so it may be preferable to use Anick's approach if the boundaries are not needed explicitly (for instance, in discussions of Euler characteristic or cohomological dimension).

Squier's construction is slightly different (and it is also necessary to interchange left and right in order to make comparisons). Squier's P_3 has as basis all overlap ambiguities u, v, w. Anick's P_3 is smaller. Its basis consists of those overlap ambiguities u, v, w for which there is no overlap ambiguity u', v', w' with $u'v'w'$ an initial segment of uvw. Also, Squier uses a more general expression for ∂_3. Squier's construction of ∂_3 depends on a choice, for each string w, of a sequence $w = w_0, w_1, ..., w_n$ such that w_n is irreducible and, for all $i < n$, either $w_i \to w_{i+1}$ or $w_{i+1} \to w_i$. If this choice is made so that w_{i+1} always comes from w_i by leftmost rewriting then this ∂_3 is the same as Anick's. It is interesting to note that the first hypothesis of Squier's Theorem 3.2 amounts to saying that his P_3 is the same as Anick's, while his second hypothesis ensures that Anick's P_4 is zero; thus his conclusion is immediate.

In an unpublished paper, Squier has shown that a monoid which can be presented by a finite complete rewriting system satisfies an additional condition, which he refers to as *having finite derivation type*, and he exhibits a finitely presented FP_∞ monoid with solvable word problem which is not of finite derivation type. I do not properly understand this criterion, but it appears to be of a homotopical nature, instead of being homological.

Squier's unpublished results, and Brown's work, give rise to a number of interesting problems.

Are monoids with solvable word problem and of finite derivation type necessarily presented by a finite complete rewriting system? (My guess is "No".)

Is a monoid of finite derivation type necessarily FP_∞?

Do automatic groups [9] (which are known to be FP_∞ with solvable word problem) have presentations by finite complete rewriting systems? Are they of finite derivation type?

There are finitely presented infinite simple groups (which of necessity have solvable word problem) which are FP_∞ [18, 27]. One of these is given explicitly in [6], in a form from which a presentation could easily be written down. Does this group (more generally, this family of groups) have a presentation by a finite complete rewriting sytem? Is it (are they) of finite derivation type?

The group $\langle x_i \ (i \in \mathbb{N}); \ x_j x_i = x_i x_{j+1} \ (\text{all } i, j \text{ with } i < j) \rangle$ is discussed in [7], where a finite presentation is given, and the group is shown to be FP_∞. It is easy to see that this group has solvable word problem. Can it be presented by a finite complete rewriting system? Does it have finite derivation type?

Let F be a free group of rank n. According to Culler and Vogtmann [12], the group $\mathrm{Out}(F)$ of outer automorphisms of F has a subgroup of finite index whose cohomological dimension is $2n - 3$. They prove the result by exhibiting a complex on which the subgroup acts. Can it be proved by applying Brown's result (or Anick's, equivalently) to some presentation?

References

[1] D.J. Anick, *On the homology of associative algebras,* Trans. Amer. Math. Soc. **296** (1986), pp. 641–659.

[2] G. Baumslag, R. Bieri, *Constructable solvable groups,* Math. Z. **151** (1976), pp. 249–257.

[3] B. Benninghoven, S. Kemmerich and M.M. Richter, *Systems of reductions,* Lecture Notes in Computer Science 277, Springer, 1987.

[4] G. Bergman, *The Diamond Lemma for ring theory,* Adv. in Math **29** (1978), pp. 178–218.

[5] K.S. Brown, *The geometry of rewriting systems; a proof of the Anick-Groves-Squier theorem,* in: Algorithms and Classification in Combinatorial Group Theory (G. Baumslag, C.F. Miller, eds.), MSRI Publications 23, Springer, New York, 1992, pp. 137–164.

[6] K.S. Brown, *The geometry of finitely presented infinite simple groups,* in: Proceedings of the Workshop on Algorithms, Word Problems, and Classification in Combinatorial Group Theory (G. Baumslag, F. Cannonito, C.F. Miller, eds.), MSRI Publications 23, Springer, New York, 1992, pp. 121–136.

[7] K.S. Brown and R. Geoghegan, *An infinite-dimensional torsion-free FP_∞ group,* Invent. Math. **77** (1984), pp. 367–381.

[8] B. Buchberger and R. Loos, *Algebraic simplification,* in: Computer algebra (B. Buchberger, J. Collins, R. Loos, eds.), Computing Supplement **4** (1982), pp. 11–43.

[9] D.B.A. Epstein, J.W. Cannon, D.F. Holt, S.V.F. Levy, M.S. Paterson, W.P. Thurston, *Word processing in groups,* Jones and Bartlett, Boston, 1992.

[10] D.E. Cohen, *A monoid which is right FP_∞ but not left FP_1,* to appear in Bull. London Math. Soc.

[11] D.E. Cohen, *Combinatorial Group Theory,* London Mathematical Society Student Texts 14, Cambridge University Press, Cambridge, 1990.

[12] M. Culler and K. Vogtmann, *Moduli of groups and automorphisms of free groups,* Invent. Math. **84** (1986), pp. 91–119.

[13] N. Dershowitz, *Termination of Rewriting,* J. Symbolic Computation **3** (1987), pp. 69–116.

[14] N. Dershowitz and J-P. Jouannard, *Notations for Rewriting*, Bulletin of the European Association for Theoretical Computer Science **43** (February, 1991), pp. 162–173.

[15] V. Diekert, *Complete semi-Thue systems for abelian groups*, Theoretical Computer Science **44** (1986), pp. 199–208.

[16] J.R.J. Groves, *Rewriting systems and homology of groups*, in: Groups — Canberra 1989 (L.G. Kovacs, ed.), Lecture Notes in Mathematics 1456, Springer, Berlin, 1991, pp. 114–141.

[17] J.R.J. Groves and G.C. Smith, *Rewriting systems and soluble groups*, to appear.

[18] G. Higman, *Finitely Presented Infinite Simple Groups*, Notes on Pure Mathematics 8, Australian National University, Canberra, 1974.

[19] G. Huet, *Confluent reduction: abstract properties and applications to term rewriting systems*, J. Ass. Computing Machinery **27** (1980), pp. 797–821.

[20] M. Jantzen, *Confluent String Rewriting*, European Association for Theoretical Computer Science Monographs 14, Springer, Berlin, 1988.

[21] D. Kapur and P. Narendran, *A finite Thue system with a decidable word problem and without an equivalent finite canonical system*, Theoretical Computer Science **35** (1985), pp. 337–344.

[22] D. Kapur and P. Narendran, *The Knuth-Bendix completion procedure and Thue systems*, SIAM J. Computing **14** (1985), pp. 952–1072.

[23] D. Knuth and P. Bendix, *Simple word problems in universal algebra*, in: Computational problems in abstract algebra (J. Leech, ed.), Pergamon, Oxford, 1970, pp. 263–297.

[24] Y. Kobayashi, *Complete rewriting systems and homology of monoid algebras*, J. Pure Appl. Algebra **65** (1990), pp. 263–275.

[25] P. Le Chenadec, *Canonical forms in finitely presented algebras*, Pitman, London, 1986.

[26] K. Madlener and F. Otto, *Commutativity in groups presented by finite Church-Rosser Thue systems*, Informatique theorique et Applications **22** (1988), pp. 93–111.

[27] R. McKenzie and R.J. Thompson, *An elementary construction of unsolvable word problems in group theory*, in: Word Problems: Irvine 1969 Conference (W.W .Boone, F.B. Cannonito, R.C. Lyndon, eds.), Studies in Logic and the Foundations of Mathematics North-Holland, 1973, pp. 457–478.

[28] M.H.A. Newman, *On theories with a combinatorial definition of equivalence*, Ann. of Math. **43** (1943), pp. 223–243.

[29] C.C. Squier, *Word problems and a homological finiteness condition for monoids*, J. Pure Appl. Algebra **49** (1987), pp. 201–217.

Added in proof.

    D.R.Farkas, *The Anick resolution*, J. Pure Appl. Algebra **79** (1992), pp. 159–168.

    S.M. Hermiller, *Rewriting systems for Coxeter groups.*

One Relator Products with High-Powered Relators

Andrew J. Duncan and James Howie

Department of Mathematics, Heriot-Watt University, Riccarton, Edinburgh, EH14 4AS.

Abstract. If $G = \langle x_1, \ldots, x_n \mid r^m \rangle$ is a one-relator group for some large integer m, then it is well known that G has many nice properties, and these are often easier to prove than in the torsion-free case (where the relator is not a proper power). To a large extent, this phenomenon also occurs for a one-relator product $G = (A * B)/N(r^m)$ of arbitrary groups A, B (where $N(\cdot)$ means normal closure). In this article we survey some recent results about such groups, describe the geometric methods used to prove these results, and discuss what happens for lower values of m. Specifically, we give counterexamples to a conjecture made in [36].

1. Introduction

Given a set $X = \{x_1, x_2, x_3, \ldots\}$ and a word r on $X \cup X^{-1}$ the group G given by the presentation $\langle x_1, x_2, x_3, \ldots \mid r \rangle$ is said to be a *one-relator* group. An extensive and successful theory of one-relator groups has been established, based largely on the work of W. Magnus, and generalising the theory of free groups (see [44], Chapter II). The theory of one-relator groups provides a basic model and an extensive stock of techniques for further generalization of free groups. With this aim in mind, one possibility is to consider presentations with more than one relator. This idea has been been successfully pursued by I.L. Anshel [1] for the case of two-relator groups (see also [40] for a particular class of two-relator groups). Bogley [4] has also extended a number of one-relator group theorems to certain groups with arbitrarily many relations.

Another possibility, that which we consider here, is to generalize from one-relator groups to one-relator products of groups. Let A and B be groups, let s be a word in the free product $A * B$ and let $N(s)$ be the normal closure of

Research partly supported by SERC grant GR/E 88998

the subgroup generated by s in $A * B$. The group $G = (A * B)/N(s)$ is then called a *one-relator product* of A and B. In this case the word s is called the *relator* and A and B the *factors*. At first sight it seems that there is little hope of any satisfactory theory passing on from one-relator groups to one-relator products. A fundamental Theorem of one-relator groups is the Freiheitssatz of Magnus which does not in general hold in the case of one-relator products (see §3). However by imposing restrictions, either on the factor groups A and B or on the relator s progress can be achieved. A restriction on the factors which seems to be just strong enough to make theorems go through is that A and B be *locally indicable* (in other words, every nontrivial, finitely generated subgroup admits a homomorphism onto the infinite cyclic group C_∞). This condition does lead to a successful theory (see [2, 7, 8, 16, 17, 20, 23, 33, 34, 35, 39, 54]). On the other hand the results presented in this article arise from the imposition of a condition on the relator.

Let r be a word and $s = r^m$, where m is a non-negative integer. Then s is an *m-th power* of r. The *root* of s is the smallest non-trivial subword r of s such that s is identical to a power of r. In the theory of one-relator groups the condition that the relator is an m-th power, for a large integer m, leads to geometric proofs that are both simpler and more illuminating than the proofs available in the general case. In particular for large m suitable small-cancellation hypotheses are satisfied, and the results of this theory can be applied. This behaviour is to a large extent mirrored by one-relator products.

The purpose of this survey is to describe the results that can be achieved when the relator in a one-relator product is an m-th power for various different values of m. As a general rule the larger m is the more can be said and the easier the proofs. In fact when $m \geq 7$ relatively simple geometric proofs can be given for almost anything that is true of one-relator groups. For values of m of 4 or less progress is much slower and in some cases the methods break down altogether. Further restrictions that can be imposed to obtain results at these low values of m are, for instance, that the relator contain no A or B letters of order 2, or that the factors be locally indicable (see for example Corollary 3.2, Theorem 3.10, Theorem 3.13 or Theorem 3.14).

Most of the results in this article are stated without proof, since they are proved elsewhere in the literature. Where a result is new, we indicate this and give at least an outline of the proof. The proofs given here and elsewhere are largely geometric, relying heavily on the analysis of pictures over groups. Pictures are the dual of van Kampen diagrams that arise in small cancellation theory. A summary of the relevant facts about pictures can be found in §2.

Historically, the first result about one-relator products with a high powered relator seems to be due to Gonzalez-Acuña and Short [29], who proved a version of the Freiheitssatz for sixth and higher powers, and applied it to the problem of which Dehn surgeries on knots in S^3 can yield reducible manifolds.

The latter problem has since been completely solved [6, 30], again using one-relator products. See §5 (1) for details.

The paper is laid out as follows. Section 2 gives a brief introduction to pictures over a one relator product, including the various ideas that appear in our proofs. In Section 3 the main results are stated, with proofs or sketches of proofs where the result is new. Section 4 discusses the breakdown at low values of m of the basic theorem on pictures, namely "Conjecture F" (which was proved for $m \geq 4$ in [36, 37] and conjectured for $m \geq 2$). Some applications of the theory (including the Dehn surgery problem mentioned above) are given in Section 5 and some open problems in Section 6.

2. Pictures

Pictures were introduced by Rourke [53] as a technique in group theory, and have been applied by several authors in this way (see [52] and the references cited there). They are essentially the dual of Dehn (van Kampen, Lyndon) diagrams, which are familiar from small cancellation theory. See [9], [25] or [52] for a full discussion of pictures.

Pictures were adapted to the context of one-relator products by Short [54], who used them to prove the Freiheitssatz for locally indicable factors. They have since been applied in this way by various authors [16, 17, 19, 29, 36, 37, 38, 40].

We shall give a brief description of the technique here, and indicate how they are applied to various problems. A more detailed introduction can be found in [36].

Let Σ be a compact surface, and $G = (A * B)/N(r^m)$ a one-relator product. A *picture* Γ on Σ over G consists of:

  (i) A disjoint union of (small) discs $v_1, \ldots, v_n$ in int(Σ) (called the *vertices* of Γ);

 (ii) A properly embedded 1-submanifold ξ of $\Sigma_0 = \Sigma \setminus \text{int}(\bigcup v_i)$ (whose components are called the *arcs* of Γ, even when they are closed curves);

(iii) An orientation of $\partial\Sigma_0$, and a *labelling function*, that associates to each component of $\partial\Sigma_0 \setminus \xi$ a *label*, which is an element of $A \cup B$.

This data is required to satisfy a number of properties.

  (a) In any *region* Δ of Γ (that is, any component of $\Sigma \setminus (\bigcup v_i \cup \xi)$), either all labels belong to A or all labels belong to B (we will refer to Δ as an A-region or a B-region accordingly);

  (b) Each arc separates an A-region from a B-region;

  (c) The *vertex label* of any vertex v_i (that is, the word consisting of the labels of ∂v_i read in the direction of its orientation from some starting point) is identically equal to r^m (in the free monoid on $A \cup B$) up to cyclic permutation;

(d) Suppose Δ is an orientable region of Γ, of genus g say, with k boundary components. Each boundary component has a *boundary label* defined as follows. If $a_1, \ldots, a_t$ are the labels of that boundary component, in the cyclic order of the orientation induced from some fixed orientation of Δ, then the boundary label is

$$a_1^{\epsilon_1} \ldots a_t^{\epsilon_t},$$

where ϵ_t is $+1$ if the orientation of the segment of $\partial \Sigma_0$ labelled a_i agrees with that of the boundary component of Δ, and -1 otherwise. If $\alpha_1, \ldots \alpha_k$ are the boundary labels of Δ, then the equation

$$X_1 \alpha_1 X_1^{-1} \ldots X_k \alpha_k X_k^{-1} [Y_1, Z_1] \ldots [Y_g, Z_g] = 1$$

is solvable (for X_i, Y_j, Z_j) in A (if Δ is an A-region) or in B (if Δ is a B-region).

There is a similar condition for nonorientable regions, where the commutators are replaced by squares in the equation (see [17]). We omit the details.

In [36] it is shown how to associate pictures to a map from Σ to a certain space with fundamental group G. For example, elements of the second homotopy group of this space can be represented by *spherical pictures*, that is by pictures on S^2. Pictures on surfaces with boundary correspond to solutions of certain quadratic equations in G, as follows. If Γ is a picture on a surface Σ over a one-relator product $G = (A * B)/N(r^m)$, and β is a boundary component of Σ, then the *label* of β is just the product of the labels of the segments contained in β, read in the order of the orientation of β.

Proposition 2.1. *Let Σ be a compact orientable surface of genus g with k boundary components, and let $G = (A * B)/N(r^m)$ a one-relator product. Suppose that $W_1, \ldots, W_k$ are cyclically reduced words in $A * B$. Then there exists a picture on Σ over G with boundary labels $W_1, \ldots, W_k$ (where the orientation of $\partial \Sigma$ is induced from that of Σ) if and only if the equation*

$$X_1 W_1 X_1^{-1} \ldots X_k W_k X_k^{-1} [Y_1, Z_1] \ldots [Y_g, Z_g] = 1$$

can be solved (for X_i, Y_j, Z_j) in G.

Indeed, given any path-connected space P with fundamental group G, the equation in the Proposition is soluble if and only if there is a continuous map from Σ to P whose restriction to $\partial \Sigma$ represents the k classes $W_1, \ldots, W_k \in \pi_1(P)$. The techniques of [36] associate a picture to any such map, for a suitable choice of P, and *vice versa*. See [16] for a discussion. There is also a version for nonorientable surfaces [17], but we omit the details. This result is particularly useful where $\Sigma = D^2$. There is a picture over G on D^2 with boundary label W if and only if $W = 1$ in G.

Example Figure 1 shows a picture on D^2 over the one-relator product $G = (A * B)/N((ab)^2)$, where $A = \langle a \mid a^3 = 1 \rangle$ and $B = \langle b \rangle$, representing the fact that $(ab^2)^{-3} = 1$ in G.

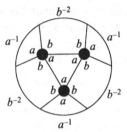

Figure 1

We allow a number of operations on pictures as follows. First of all, a *bridge move* can be performed along an embedded curve γ in Σ meeting Γ only in its endpoints, which must be interior points of arcs of Γ, under suitable conditions.

The interior of γ is necessarily contained in some region Δ of Γ. A bridge move is essentially surgery on ξ in a neighbourhood of γ. Consider a thin rectangular neighbourhood of γ, whose short edges are contained in the arcs of Γ. Simply replace these short edges by the two long edges of the rectangle. We permit this move, and call it a bridge move, provided the result satisfies the rules for a picture. See Figure 2.

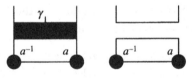

Figure 2

A spherical picture containing precisely two vertices is called a *dipole*. We also use this term to denote a two-vertex picture on D^2 with no arcs meeting ∂D^2. We allow ourselves to insert or delete 'floating' dipoles. That is, if $D \subset \Sigma$ is a disc such that Γ restricts to a dipole Γ_0 on D, we may replace Γ_0 by the empty picture on D, and *vice versa*.

We say that a pair of vertices of a picture Γ *cancel* along an arc α if they are joined by α and if they can be made into a dipole using bridge moves on the incident arcs other than α.

Suppose Δ is a disc-region of Γ, whose boundary consists of two arcs and two segments of $\partial \Sigma_0$. Then we say that the two arcs in $\partial \Delta$ are *parallel*. The two segments of $\partial \Sigma_0$ in $\partial \Delta$ are labelled either by identical letters occurring in r or by mutually inverse letters occurring in r, according to their relative

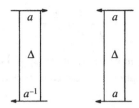

Figure 3

orientations (see Figure 3). Moreover, in the former case, unless the two vertices concerned cancel, the two letters concerned must occur at different places in r.

We extend the notion of parallelism of arcs to an equivalence relation in the obvious way. If we have an equivalence class containing t arcs, then there are $t-1$ regions separating them, and the result is that we have two sequences of letters representing identical subwords of r^m (or of r^m and r^{-m}). Moreover, in the first case the two subwords occur in distinct places in r^m, unless the two vertices concerned cancel.

If Γ is *reduced* (that is, no pair of vertices cancel), then it turns out that the upper bound for the size of a parallelism class of arcs is $\ell = \text{length}(r)$ [36]. Moreover, this upper bound can be attained only if r has the form (up to cyclic permutation) $xUyU^{-1}$ for some word U and some letters x, y with $x^2 \neq 1$, $y^2 = 1$. The subword UyU^{-1} is then identical to its own inverse, allowing a class of ℓ arcs (see Figure 4).

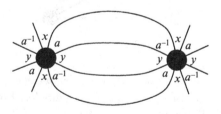

Figure 4

In view of the above, it turns out that some special forms of r are worth special attention. We say that r is *exceptional* if it has the form

$$xUyU^{-1}$$

for some word U and letters x, y (up to cyclic permutation). If p, q are the orders of x, y respectively, we say that G is of type $E(p, q, m)$. In the case where U is empty, $A = \langle x \rangle$ and $B = \langle y \rangle$, G is then the triangle group of type (p, q, m), which we denote $G_0(p, q, m)$. If in addition the number

$$s = \frac{1}{p} + \frac{1}{q} + \frac{1}{m} - 1$$

is positive, then $G_0(p,q,m)$ is finite of order $2/s$. For a finite triangle group $G = G_0(p,q,m)$, there is a canonical spherical picture $\Gamma(p,q,m)$ arising from

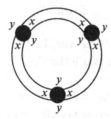

Figure 5

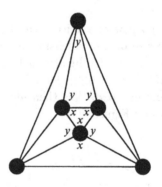

Figure 6

the action of G on S^2. It has $2/sm$ vertices, $2/sp$ p-sided A-regions, and $2/sq$ q-sided B-regions. (In particular, $\Gamma(2,2,m)$ is a dipole.) Figures 5 and 6 show $\Gamma(2,3,2)$ and $\Gamma(3,3,2)$ respectively. Illustrations of $\Gamma(2,3,m)$, $m = 3,4,5$, can be found in [36].

In general, if G is exceptional, of type $E(p,q,m)$ and $s > 0$, then there is a natural homomorphism from $G_0(p,q,m)$ to G, and $\Gamma(p,q,m)$ induces a spherical picture over G, which we also call $\Gamma(p,q,m)$ by abuse of notation. The final operation we allow on pictures is the insertion or deletion of a floating $\Gamma(p,q,m)$.

Note that it is possible for a word r, and so a one-relator product G, to be exceptional in more than one way, possibly with different values of p,q,s. For example if a,b denote letters of A,B respectively of orders 2 and 3 respectively, then $(abab^2)^m$ is both of type $E(2,2,m)$ and of type $E(3,3,m)$, while $(ababab^2)^m$ is of type $E(2,3,m)$ in two distinct ways. This phenomenon played a part in [37], where the notation $E(2,3,4)^+$ was introduced to denote a relator that is exceptional of type $E(2,3,4)$ in two different ways. In general, we will say that r (or G) is *uniquely exceptional* if it can be regarded as exceptional in only one way, and *multiply exceptional* otherwise. For multiply exceptional words, there will be more than one natural homomorphism $G_0(p,q,m) \rightarrow G$ (possibly for different values of p,q), and more than one natural spherical picture $\Gamma(p,q,m)$ over G. A great deal of care is required when dealing with this situation, as will become apparent later.

Two pictures over G (on the same surface Σ) are said to be *equivalent* if each can be obtained from the other by a sequence of allowable moves, namely bridge moves and insertion/deletion of floating dipoles or $\Gamma(p,q,m)$'s. Maps on Σ represented by equivalent pictures differ up to homotopy only by an element of the $\mathbb{Z}G$-submodule of π_2 generated by the classes of dipoles and $\Gamma(p,q,m)$'s. In particular, if Σ has a boundary, then the boundary labels of equivalent pictures on Σ are equal (up to cyclic permutation).

We say that a picture over G on Σ is *efficient* if it has the least number of vertices in its equivalence class. In particular, efficient pictures are always reduced.

Given an efficient picture Γ on a surface Σ, we associate *angles* to the various components of $\partial\Sigma_0 \smallsetminus \xi$ and use these to compute *curvatures* for each region and vertex of Γ. There are various possible ways of associating angles, but once that has been done, we define curvature as follows. If v is a vertex and $\sigma(v)$ is the sum of the angles at v, we define the *curvature* of v to be $\kappa(v) = 2\pi - \sigma(v)$. If β is a boundary component of Σ, we define the *curvature* of β to be $\kappa(\beta) = -\sigma(\beta)$, where $\sigma(\beta)$ is the sum of the angles on β. Finally, if Δ is a region of Γ, and $\sigma(\Delta)$ is the sum of the angles in Δ, we define the *curvature* of Δ to be $\kappa(\Delta) = \sigma(\Delta) + 2\pi\chi(\Delta) - n\pi$, where $\chi(\Delta)$ is the Euler characteristic of Δ, and n is the number of arcs in $\partial\Delta$ that are not closed.

The sum of all these curvatures, over all vertices, boundary components and regions of Γ, is then just $2\pi\chi(\Sigma)$. The general idea is to assign angles in such a way that positive curvature for vertices, regions or boundary components is rare (or better still, to get negative upper bounds). This will then limit the pictures that can actually arise.

Before proceeding, it should perhaps be noted that the approach to curvature described above is slightly different from those used in [16, 17, 36, 37, 38], particularly regarding boundary components of Σ. However, all these approaches are essentially equivalent.

A region of Γ is a *boundary region* if it meets $\partial\Sigma$. It is a *simple boundary region* if it meets $\partial\Sigma$ in a single segment. A vertex v of Γ is a *boundary vertex* if it is incident at a boundary region, and an *interior vertex* otherwise. The following are the main restrictions on curvature for pictures.

Theorem 2.2. [16], [17] *Let $G = (A * B)/N(r^m)$ be a one-relator product of two locally indicable groups A, B, and let Γ be an efficient picture over G on some compact surface Σ. Assume that each component of Σ contains part of Γ. Then there is a non-negative angle assignment such that every vertex of Γ has curvature at most $2\pi(1-m)$, every boundary component of Σ has curvature 0, and any positively curved region is either a simple boundary region or a disc bounded by a closed arc of Γ.*

If $m \geq 2$, this leads to a bound for the number of vertices in an efficient picture on a given surface with given boundary labels, which in turn leads to solutions for the genus problems for G [16], [17], see §3. With $\Sigma = D^2$ it shows that a nonempty efficient picture must have arcs meeting ∂D^2, which yields a proof of the Freiheitssatz for locally indicable factor groups.

Theorem 2.3. [29], [36], [37] *Let $G = (A * B)/N(r^m)$ be a one-relator product with $m \geq 4$, and let Γ be an efficient picture over G on a compact surface Σ. Then there is a non-negative angle assignment such that all regions have curvature zero, and unless G is of type $E(2,3,4)$ or $E(2,3,5)$ there are no positively curved interior vertices.*

This forces any positive curvature to be concentrated near the boundary of Σ. In the cases $E(2,3,4)$ and $E(2,3,5)$ it is possible to have positively curved interior vertices, but the possible configurations are very restricted. In [36], [37] this is exploited to trade the positive curvature of such vertices off against negative curvature of neighbouring vertices, which again forces any overall positive curvature to be concentrated near $\partial\Sigma$.

Theorem 2.4. *Let $G = (A * B)/N(r^3)$ be a one-relator product such that no letter occurring in r has order 2 in A or B, and let Γ be an efficient picture over G on a compact surface Σ. Then there is a non-negative angle assignment for which all regions have curvature 0, and there are no interior positively curved vertices.*

In order to prove that some one-relator product $G = (A * B)/N(r^m)$ satisfies the Freiheitssatz, for example, one would like to show that every nonempty efficient picture over G on D^2 has arcs meeting ∂D^2. In practice, it is often easier to prove a stronger result, more suited to inductive arguments. The following conjecture was made in [36] with this in mind.

Conjecture F. *Let $G = (A * B)/N(r^m)$ be a one-relator product, where r is cyclically reduced of length at least 2, and $m \geq 2$. Let Γ be an efficient picture on D^2 over G, such that at most 3 vertices of Γ are connected to ∂D^2 by arcs. Then every vertex of Γ is connected to ∂D^2 by arcs.*

The Freiheitssatz and various other results can be deduced for any one-relator product that satisfies Conjecture F (see §3). Unfortunately, Conjecture F turns out to be false in general (§4), but it does hold in many special cases.

Theorem 2.5. *Suppose* $G = (A * B)/N(r^m)$ *is a one-relator product with r cyclically reduced of length at least 2. If at least one of the following conditions hold then G satisfies Conjecture F.*

(i) $m \geq 4$;

(ii) $m = 3$ *and no letter occurring in r has order 2;*

(iii) $m = 2$ *and* A, B *are locally indicable.*

Part (i) of this theorem was proved in [36] in the case $m \geq 5$, and in [37] for $m = 4$. Part (ii) was proved in [18]. The curvature results above, Theorems 2.3 and 2.4 are crucial elements in the proofs. Here we sketch a proof of part (iii), using Theorem 2.2.

Let Γ be an efficient picture over G on D^2, with at most 3 vertices connected by arcs to ∂D^2. We may clearly assume that Γ has at least one vertex, and that no arc of Γ is either a closed curve or joins ∂D^2 to ∂D^2. Indeed, an easy induction allows us to assume that Γ is connected, that no arc joins any vertex to itself, and that any two arcs joining the same two vertices are parallel.

Applying Theorem 2.2, we have an angle assignment for which every vertex has curvature at most -2π, ∂D^2 has curvature 0, and the only positively curved regions are simple boundary regions. In fact, adapting the method of [16] slightly, this can be done in such a way that the angles are all either 0 or π, and those on components of $\partial D^2 \smallsetminus \xi$ are all 0. We amend this angle assignment slightly as follows. Given any simple boundary region Δ, there is by hypothesis at least one vertex v joined to ∂D^2 by an arc in $\partial \Delta$. We change the angle of the segment of ∂v in $\partial \Delta$ adjacent to that arc to 0. From the argument in the proof of [16], Theorem 3.3, it is clear that this makes $\kappa(\Delta)$ nonpositive.

Repeating this for all simple boundary components, we transfer all the positive curvature to vertices connected to ∂D^2. Moreover, since no angles are negative, such vertices have curvature at most $+2\pi$. Since there are at most three such vertices, and the total curvature is 2π, there are at most five vertices in all.

If u is a vertex not connected to ∂D^2, then it is known that at least $2m = 4$ parallelism classes of arcs are incident at u ([16], Lemma 3.1). Since no arc joins u to itself, and no two non-parallel arcs join u to the same neighbouring vertex, we must have at least five vertices.

We are now reduced to the case where Γ has precisely three vertices v_1, v_2, v_3 connected to ∂D^2, and two vertices u_1, u_2 not connected to ∂D^2. By the above, each of v_1, v_2, v_3 is connected to each of $u_1, u_2, \partial D^2$, contradicting the fact that the complete bipartite graph $K_{3,3}$ is non-planar.

3. Main results

In cases where Conjecture F holds, the way to generalise one-relator group theory to one-relator products is open. In practice this means that m must be sufficiently large, as is apparent in the light of the previous section.

As a first example consider the Freiheitssatz of Magnus [45]: if A and B are free groups then A and B embed in $(A * B)/N(s)$, whenever s is a non-trivial cyclically reduced word of length at least 2. For one-relator products the analogous property is the following.

The Freiheitssatz. *Let $G = (A * B)/N(r^m)$ where r is a cyclically reduced non-trivial word of length at least 2 in $A * B$ and is not a proper power. Then the Freiheitssatz holds for G if the natural maps from A and B into G are injections.*

Theorem 3.1. [36], [37] *Assume that Conjecture F holds for pictures over G on the disk. Then the Freiheitssatz holds for G.*

Proof. Suppose $a \in (A \cap N(r^m))$. Choose CW-complexes X and Y with $\pi_1(X) = A$ and $\pi_1(Y) = B$ and join X and Y with a 1-cell e^1, to obtain a space with fundamental group $A * B$. Let $Z = X \cup Y \cup e^1 \cup_{r^m} e^2$ be the space obtained by attaching a 2-cell e^2 along a path representing r^m. Let $f : S^1 \longrightarrow Z$ map S^1 homeomorphically onto a path representing a. Then f can be extended to D^2, since a is trivial in G, to give a picture Γ over G on D^2 with boundary label a. We may assume that Γ is efficient, so has no arcs meeting the boundary ∂D^2, and so by Conjecture F has no vertices at all. Hence $a = 1_A$. $\quad\blacksquare$

From the results of Section 2 we obtain:

Corollary 3.2. [29], [36], [37], [18] *Assume that either $m \geq 4$ or $m \geq 3$ and r contains no letters of order 2. Then the Freiheitssatz holds.*

The conditions of the above Corollary cannot simply be removed as, for example, $A * B = N(ab)$ when $A = C_2, B = C_3$ and a and b are non-trivial elements of A and B, respectively. However there are examples, with $m = 2$ or 3, for which the Freiheitssatz holds although Conjecture F fails (see Section 4). On the other hand there are other versions of the Freiheitssatz under conditions on A and B. The first concerns generalized triangle groups.

Theorem 3.3. *Let $G = (A * B)/N(r^m)$ where each of A and B is a finite cyclic group. Then the Freiheitssatz holds for G and r has order m in G.*

This was proved by Boyer [6] and independently by Baumslag, Morgan and Shalen [3]. The proofs work by studying representations of G into $SO(3)$ and $PSL(2, \mathbb{C})$ respectively. In fact, a more general version is true [26].

Theorem 3.4. *Let $G = (A * B)/N(r^m)$ where each of A and B is isomorphic to a subgroup of $PSL(2, \mathbb{C})$. Then the Freiheitssatz holds for G and r has order m in G.*

The proof gives a representation $\rho : G \to PSL(2, \mathbb{C})$ such that the induced representations on A and B are faithful, and $\rho(r)$ has order m.

Finally, if A and B are locally indicable then the Freiheitssatz holds for arbitrary m [8, 33, 54]. This last result gives a full generalization of Magnus' Freiheitssatz: whether or not 'locally indicable' can be replaced here by 'torsion free' remains unknown.

Using a similar argument a generalization of a theorem of Weinbaum [57] can be obtained.

Theorem 3.5. *Let $G = (A * B)/N(r^m)$ where r is a cyclically reduced non-trivial word of length at least 2 in $A * B$ and is not a proper power. If Conjecture F holds for G then no proper cyclic subword of r^m represents the identity in G. In particular r represents an element of order m in G.*

A similar result also holds when $m = 1$ if A, B are locally indicable [34]. Furthermore, using the Freiheitssatz, the following can be proved [36], [37]. This generalizes Lyndon's Identity Theorem [43].

Theorem 3.6. *Let $G = (A * B)/N(r^m)$ where r is a cyclically reduced non-trivial word of length at least 2 in $A * B$ which is non-exceptional and let N denote $N(r^m)$. If Conjecture F holds for G then N^{ab} is isomorphic as a $\mathbb{Z}G$-module to $\mathbb{Z}G/(1-r)\mathbb{Z}G$.*

Corollary 3.7. *Under the hypotheses of the theorem, the restriction-induced maps*

$$H^t(G; -) \to H^t(A; -) \times H^t(B; -) \times H^t(C; -)$$

are natural isomorphisms of functors on $\mathbb{Z}G$-modules for $t \geq 3$, and a natural epimorphism for $t = 2$, where $C = \langle r \rangle$ is the cyclic subgroup of order m generated by r. Dually, the maps

$$H_t(G; -) \leftarrow H_t(A; -) \oplus H_t(B; -) \oplus H_t(C; -)$$

are natural isomorphisms for $t \geq 3$ and a natural monomorphism for $t = 2$.

Applying a theorem of Serre [42], we then have

Corollary 3.8. *In the situation of the theorem, if $K \neq \{1\}$ is a finite subgroup of G, then $K \subset gAg^{-1}$, $K \subset gBg^{-1}$ or $K \subset gCg^{-1}$ for some $g \in G$. Moreover, precisely one of these occurs, and the left coset gA (respectively gB, gC) is uniquely determined by K.*

In particular, $A \cap C = B \cap C = \{1\}$, and $A \cap B$ is torsion free. In fact, it follows immediately from Conjecture F that $A \cap gBg^{-1} = \{1\}$ for all $g \in G$.

Similar results to the above also hold for $m = 1$ if A, B are locally indicable [35], although it is no longer necessarily true that $A \cap B = \{1\}$. In this case, since A and B are torsion-free, every finite subgroup of G is cyclic, generated by a conjugate of a power of r.

If r is exceptional (with $s > 0$), and Conjecture F holds for G, then it is not too difficult to see that the natural map $G_0(p, q, m) \to G$ is injective [36]. This holds, for example, if $r = (xUyU^{-1})^4$, where x, y have orders 2 and 3 respectively. In particular we may identify G_0 with a finite subgroup of G that properly contains C, as well as having nontrivial intersections with conjugates of A and/or B. Hence the above corollary, and so also Theorem 3.6 and Corollary 3.7, fails in this case. The main reason for the difference is the existence of the nontrivial spherical diagram $\Gamma(p, q, m)$, so that we cannot construct a $K(G, 1)$-space for G in exactly the same way as we can in the non-exceptional case. It follows that the cohomology calculation is different. However, from the fact that all efficient spherical pictures are empty, we can still deduce the following.

Theorem 3.9. *Let* $G = (A * B)/N(r^m)$ *be a one-relator product with* $m \geq 4$, *where* r *has a unique exceptional form* $E(p, q, m)$ *for which* $s > 0$. *Then the push-out of groups:*

$$
\begin{array}{ccc}
C_p * C_q & \longrightarrow & A * B \\
\downarrow & & \downarrow \\
G_0(p, q, m) & \longrightarrow & G
\end{array}
$$

induces a Mayer-Vietoris sequence of cohomology functors:

$$\cdots \to H^t(G; -) \to H^t(A*B; -) \times H^t(G_0(p, q, m); -) \to H^t(C_p * C_q; -) \to \cdots$$

Note that multiply exceptional relator words are allowed here, but only if all but one exceptional form has $s \leq 0$. Thus, for example, if a, b are letters of orders 2, 3 respectively, then $(abab^2)^4$ is allowed; it has one exceptional form $E(2, 2, 4)$ $(s = \frac{1}{4})$ and one form $E(3, 3, 4)$ $(s = -\frac{1}{12})$. On the other hand $(ababab^2)^4$ is not allowed, since it has two distinct exceptional forms of type $E(2, 3, 4)$. The theorem was (wrongly) stated in [36] without any restrictions on multiply exceptional words, but the proof does not work in that generality.

The proof works by showing that the pushout diagram can be realized as the fundamental groups of an adjunction-space diagram of aspherical spaces [36]. There is also a dual Mayer-Vietoris sequence of homology functors.

There are many other examples of standard results of one-relator group theory that have natural generalizations to one-relator products. The next three theorems are generalizations of the Spelling Theorems of Newman and Gurevich [31], [50].

Theorem 3.10. *Let r be a cyclically reduced word of length $l \geq 2$ in the free product $A * B$. Assume that $m \geq 4$ and that r^m is not of the form $E(2,3,4)$ or $E(2,3,5)$. Let W be a non-empty, cyclically reduced word belonging to the normal closure of r^m. Then either*

(1) W is a cyclic permutation of $r^{\pm m}$; or

(2) W has two disjoint cyclic subwords U_1, U_2 such that each U_i is identical to a cyclic subword V_i of $r^{\pm m}$, $l(U_1) = l(U_2) \geq (m-1)l - 1$ and W has a cyclic permutation $x_1 U_1 x_2 U_2$ for some elements x_1, x_2 of the pregroup $A \cup B$; or

(3) W has k disjoint cyclic subwords $U_1, \ldots, U_k$ for some $k \in \{3,4,5,6\}$ such that each U_i is identical to a cyclic subword V_i of $r^{\pm m}$, and V_i has length at least $(m-2)l - 1$ for $i \leq 6 - k$, and at least $(m-3)l - 1$ for $i > 6 - k$.

Theorem 3.11. *Let r be a cyclically reduced word of length $l \geq 2$ in the free product $A * B$, such that no letter of r has order 2, and let $m \geq 3$. Let W be a non-empty, cyclically reduced word belonging to the normal closure of r^m. Then either*

(1) W is a cyclic permutation of $r^{\pm m}$; or

(2) W has two disjoint cyclic subwords U_1, U_2 such that each U_i is identical to a cyclic subword V_i of $r^{\pm m}$, $l(U_1) = l(U_2) \geq (m-1)l$ and W has a cyclic permutation of the form $x_1 U_1 x_2 U_2$ for some elements x_1, x_2 of the pregroup $A \cup B$; or

(3) W has three disjoint cyclic subwords U_1, U_2, U_3 such that each U_i is identical to a cyclic subword V_i of $r^{\pm m}$ of length at least $(m-2)l$.

Theorem 3.12. *Let r be a cyclically reduced word of length $l \geq 2$ in the free product $A * B$. Assume that A and B are locally indicable. Let W be a non-empty, cyclically reduced word belonging to the normal closure of r^m. Then either*

(1) W is a cyclic permutation of $r^{\pm m}$; or

(2) W has two disjoint cyclic subwords U_1, U_2 such that each U_i is identical to a cyclic subword V_i of $r^{\pm m}$ of length at least $(m-1)l + 1$.

Theorems 3.10 and 3.11 were proved in [19], using pictures and curvature arguments as described in §2. Theorem 3.12 was proved in [39], using the dual of pictures (diagrams), and the strong properties of locally indicable groups.

Another classical result from one-relator theory is the theorem of Cohen and Lyndon [11], that the normal closure of the relator (in the underlying free group) is freely generated by a suitable collection of conjugates of the relator. This also generalizes to one-relator products under various conditions.

Theorem 3.13 *Assume that one of the following conditions hold:*

(i) $m \geq 6$;

(ii) $m \geq 4$ and no letter of r has order 2;

(iii) A and B are locally indicable.

Then there is a set U of double coset representatives of $N(r^m)\backslash(A*B)/C$ such that $N(r^m)$ is freely generated by $\{ur^m u^{-1} \mid u \in U\}$. Here C is the cyclic group generated by r, except in the case $E(2,2,m)$, $r \equiv xWyW^{-1}$, $x^2 = y^2 = 1$, in which case it is the dihedral group generated by x and WyW^{-1}.

See [7, 19, 23]. Another example is the following, which generalizes a theorem of Magnus [46].

Theorem 3.14 *Suppose r_1 and r_2 are cyclically reduced non-trivial words of length at least 2 in $A*B$, m_1 and m_2 are positive integers, and that for each i at least one of the following conditions holds:*

(i) $m_i \geq 4$ and $r_i^{m_i}$ is not of the form $E(2,3,4)$ or $E(2,3,5)$;

(ii) $m_i \geq 3$ and no letter of r_i has order 2;

(iii) A and B are locally indicable.

If $N(r_1^{m_1}) = N(r_2^{m_2})$ then $m_1 = m_2$ and r_2 is a cyclic permutation of $r_1^{\pm 1}$.

See [7, 19, 20].

We end this section with a discussion of decision problems for one-relator products. The classic decision problem is the word problem: given a word in the generators of some group, can one decide algorithmically whether or not that word represents the identity element? Magnus [47] proved that the word problem for one-relator groups is soluble. Indeed, a somewhat stronger result holds: if G is a one-relator group, and F the subgroup generated by some subset of the generators of G, then the *generalized word problem* for F in G is soluble. In other words, given any word W in the generators of G, one can decide whether or not W represents an element of F, and if so one can find a word in the generators of F that represents the same element as W.

Under suitable conditions, we can generalize the above result to one-relator products of groups.

Theorem 3.15 *Suppose that $G = (A*B)/N(r^m)$ is a one-relator product, where A and B are groups given by recursive presentations with soluble word problem, r is a cyclically reduced word in $A*B$ of length at least 2, given explicitly in terms of the generators of A and B, and that one of the following conditions holds:*

(i) $m \geq 4$;

(ii) $m = 3$ and no letter occurring in r has order 2;

(iii) $m = 2$ and A, B are locally indicable;

(iv) A, B are effectively locally indicable.

Then the generalized word problem for A (respectively, B) in G is soluble.

Parts (i) and (ii) were proved in [18], [38] (using an isoperimetric inequality derived from the curvature results of §2), and parts (iii) and (iv) by Mazurovskiĭ [49]. The term *effectively locally indicable* in (iv) means that there exists an algorithm to decide, given a finite set of words in the generators, whether or not the subgroup H of G they generate is trivial, and if not to find a homomorphism from H to C_∞. Part (iii) was also proved in [39] (using the spelling theorem), and is a special case of Theorem 3.16 below.

The second commonly studied decision problem is the conjugacy problem: to decide whether two words in the generators represent conjugate elements of a group. Another, less familiar one is the commutator recognition problem: decide whether a given word represents a commutator in the group. These are subsumed in a class of problems we call the *genus problems*. The genus problem $GP(g, n)$ is the following algorithmic problem: given n words $W_1, \ldots, W_n$ in the generators of a group, decide whether or not the equation

$$X_1 W_1 X_1^{-1} \ldots X_n W_n X_n^{-1} [Y_1, Z_1] \ldots [Y_g, Z_g] = 1$$

can be solved (for X_i, Y_j, Z_j) in the group, and if so, find an explicit solution in terms of the generators.

Thus $GP(0, 1)$ is the word problem, $GP(0, 2)$ is the conjugacy problem, and $GP(1, 1)$ is the commutator recognition problem. The problems $GP(0, n)$ were called the *dependence problems* by Pride [51]. We shall say that the genus problem is soluble for a given group if $GP(g, n)$ is soluble for all g, n. The genus problem is known to be soluble for certain small cancellation groups [51] and for negatively curved (or hyperbolic) groups [55]. It is also soluble for a free product of groups if it is soluble in each of the free factors [12, 14, 28, 58]. The conjugacy problem for one-relator groups with torsion was solved by Newman [50], and for general one-relator groups by Juhàsz (unpublished). The situation for one-relator products is as follows.

Theorem 3.16 *Let $G = (A * B)/N(r^m)$ be a one-relator product, where A, B are groups given by recursive presentations for which $GP(g', n')$ is soluble for all pairs of integers (g', n') such that $0 \le g' \le g$ and $1 \le n' \le n + 2(g - g')$, and r is a cyclically reduced word of length at least 2, given explicitly in terms of the generators of A, B. Suppose that one of the following conditions hold:*
  (i) $m \ge 5$ and G is not of type $E(2, 3, 5)$ or $(2, 3, 6)$;
  (ii) $m \ge 4$ and no letter occurring in r has order 2;
  (iii) $m \ge 2$ and A, B are locally indicable.
Then $GP(g, n)$ is soluble for G.

Corollary 3.17. *If $G = (A * B)/N(r^m)$ where A and B are groups having presentations with soluble genus problem, and one of the conditions (i), (ii), (iii) of the theorem hold, then G has soluble genus problem.*

Corollary 3.18. *The genus problem is soluble for one-relator groups with torsion.*

Part (iii) of the theorem was proved in [16]. The proofs of parts (i) and (ii) are similar, and we sketch the general argument here.

Given words $W_1, \ldots, W_n$ in the generators of A and B, we need to decide whether or not there is a picture Γ over G, on a compact orientable surface Σ of genus g with n boundary components, such that the boundary labels of Σ (with respect to the orientation of $\partial\Sigma$ induced from some orientation of Σ) are $W_1, \ldots, W_n$ (see Proposition 2.1).

Such a picture Γ may clearly be taken to be efficient, and contain no arcs which are closed curves bounding discs in Σ, and no arcs going from $\partial\Sigma$ to $\partial\Sigma$ that together with subarcs of $\partial\Sigma$ bound discs in Σ. Under any one of the hypotheses (i)-(iii) of the theorem, we may assign angles in such a way that all boundary components have curvature 0, all vertices have curvature at most $-\epsilon$, and the only positively curved regions are simple boundary regions of curvature at most $+\eta$, for some real constants $\epsilon, \eta > 0$.

The number of simple boundary regions is bounded by the number of points of $\partial\Sigma$ meeting arcs of Γ, which is the sum of the free product lengths of the W_i. This bound, together with the formula $2\pi\chi(\Sigma)$ for the total curvature of Γ, yields a bound (an *isoperimetric inequality*) for the number of vertices in Γ. Thus there are only a finite number of possible candidates for Γ (up to ambient isotopy in Σ). In order to test one of these candidates for the conditions for a picture, we apply the solutions to $GP(g', n')$ in A, B to the regions (which are subsurfaces of Σ of genus at most g and Euler characteristic at least $\chi(\Sigma)$).

The finitely many candidates for Γ can be effectively listed, and each can be effectively tested to decide whether it is a picture. Hence the problem $GP(g, n)$ is algorithmically soluble for G.

There is a nonorientable version of the genus problem [17], involving pictures on nonorientable surfaces. There are also analogous results for these problems, but there are some extra technicalities involved, arising from loss of orientability. We omit the details for ease of exposition.

4. Counterexamples to conjecture F

In general, Conjecture F fails in the cases $m = 2$ and $m = 3$. This can be seen quite easily from Figures 7 and 8, where r has the form $xUyU^{-1}$ with x of order 2 and y of order greater than 3. Two possible ways around this problem suggest themselves. Firstly, is Conjecture F stronger than is needed to obtain the main results of §3? This may well be true. However, we shall see below that some of the main results also fail for $m = 2$ and $m = 3$. A second approach would be to try to find out exactly for which cases Conjecture F fails, or at least to try to find special cases where Conjecture F remains true, with $m = 2$ or $m = 3$. Theorem 2.5 (ii) is one result of this type.

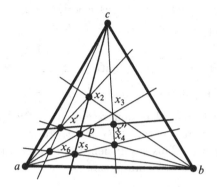

Figure 7.

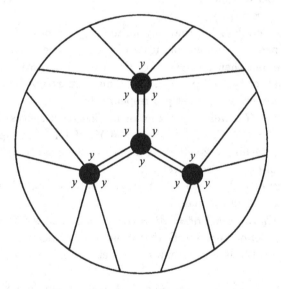

Figure 8.

Suppose that r has the form $E(p, q, m)$ with $m = 2$ or $m = 3$, and $s = 1/m + 1/p + 1/q > 1$. There is then a homomorphism to G from the finite group $G(p, q, m) = \langle x, y \mid x^p = y^q = (xy)^m = 1 \rangle$, which in all known cases is injective. In some cases of course, $G = G(p, q, m)$, so G is also a finite group. However, there are some other known examples of generalized triangle groups (with $A \cong C_p$ and $B \cong C_q$) such that G is finite of order strictly greater than $G(p, q, m)$.

Theorem 4.1. *Let $G = (A * B)/N(r^m)$ be a one-relator product of two groups A, B, such that G is finite. Assume that each of A and B embeds into G via the natural map, that r has order m in G, and that there is no nonempty efficient spherical picture over G. Then G is exceptional (of type $E(p, q, m)$, say). If in addition r^m has a unique exceptional form with the property that $s = 1/p + 1/q + 1/m - 1 > 0$, then*

$$\frac{1}{|A|} + \frac{1}{|B|} - \frac{1}{|G|} = \frac{1}{p} + \frac{1}{q} - \frac{1}{|G_0|} = \frac{1}{2} - \frac{1}{2m} + \frac{1}{2p} + \frac{1}{2q},$$

where G_0 is the finite group $G(p, q, m) = \langle x, y \mid x^p = y^q = (xy)^m = 1 \rangle$.

Proof. Assume first that G is non-exceptional. By hypothesis the Freiheitssatz holds for G, r has order m in G, and there are no non-empty efficient spherical pictures over G. This is sufficient to prove the Identity Theorem for G, and hence obtain natural isomorphisms

$$H^t(G; -) \to H^t(A; -) \times H^t(B; -) \times H^t(C_m; -)$$

of functors on $\mathbb{Z}G$-modules for each $t \geq 3$, where C_m is the cyclic subgroup generated by r (see §3).

It follows from Corollary 3.8 that any finite subgroup of G, in particular G itself, is contained in a unique conjugate of precisely one of A, B, C_m. But if $G \subset A$, for example, then $B = B \cap G \subset B \cap A = \{1\}$, a contradiction. Hence G is exceptional, of type $E(p, q, m)$, say. Thus $r \equiv xUyU^{-1}$ for some word U, where x, y are letters of order p, q respectively.

Suppose now that G is uniquely exceptional. Construct spaces W, X, Y, Z as follows. Let W be a $K(C_p * C_q, 1)$-space and Y a $K(A * B, 1)$-space. Obtain X from W by attaching a 2-cell along a path in the class $(\alpha\beta)^m$, where α, β are generators of C_p, C_q respectively. Let $\phi : W \to Y$ be a map sending α, β to x, UyU^{-1} respectively. Then $Z = Y \cup_\phi X$ is obtained from Y by attaching a 2-cell along a path in the class r^m.

Now $\pi_1(X) = G_0 = \langle \alpha, \beta \mid \alpha^p = \beta^q = (\alpha\beta)^m = 1 \rangle$ and $\pi_1(Z) = G$. It was shown in [36], proof of Theorem E, that ϕ induces an injection from G_0 to G, so we may regard G_0 as a subgroup of G. Let $n = |G|$, $n_0 = |G_0|$, $a = |A|$, and $b = |B|$.

Let $\tilde{Z}, \tilde{X}$ denote the universal covers of Z, X respectively, and $\tilde{Y}, \tilde{W}$ the subcomplexes covering Y, W. Now $\tilde{W}$ is a $K(F_0, 1)$-space, where $F_0 = \text{Ker}(C_p * C_q \to G_0)$ is a free group (by the Kuroš subgroup theorem) of rank $\rho_0 = n_0(1 - 1/p - 1/q) + 1$ (by an Euler characteristic calculation). Similarly $\tilde{Y} = K(F, 1)$, where F is free of rank $\rho = n(1 - 1/a - 1/b) + 1$.

Let $\mathcal{C}, \mathcal{D}$ be the long exact homology sequences of the pairs $(\tilde{X}, \tilde{W})$ and $(\tilde{Z}, \tilde{Y})$ respectively. Then ϕ induces a chain map $\phi_* : \mathcal{C} \otimes_{\mathbb{Z}G_0} \mathbb{Z}G \to \mathcal{D}$. Moreover $\phi_* : H_2(\tilde{X}, \tilde{W}) \otimes \mathbb{Z}G \to H_2(\tilde{Z}, \tilde{Y})$ is an isomorphism, since each term is a free module of rank one, generated by the class of the 2-cell in $X \backslash W = Z \backslash Y$.

Since $H_2(\tilde{W}) = H_1(\tilde{X}) = H_2(\tilde{Y}) = H_1(\tilde{Z}) = 0$, we have a commutative diagram with exact rows:

$$
\begin{array}{ccccccccc}
0 & \longrightarrow & H_2(\tilde{X}) \otimes \mathbb{Z}G & \longrightarrow & H_2(\tilde{X}, \tilde{W}) \otimes \mathbb{Z}G & \longrightarrow & H_1(\tilde{W}) \otimes \mathbb{Z}G & \longrightarrow & 0 \\
& & \downarrow & & \downarrow & & \downarrow & & \\
0 & \longrightarrow & H_2(\tilde{Z}) & \longrightarrow & H_2(\tilde{Z}, \tilde{Y}) & \longrightarrow & H_1(\tilde{Y}) & \longrightarrow & 0
\end{array}
$$

in which the middle vertical map is an isomorphism. Hence the left hand vertical map is injective. Moreover, the fact that r is uniquely exceptional means that there is essentially only one spherical diagram of type $\Gamma(-, -, m)$ over G (up to spherical symmetries), and so to say that there are no nonempty efficient spherical pictures over G is the same as saying that the left hand vertical map is also surjective. But an easy computation shows that the $\mathbb{Z}$-ranks of $H_2(\tilde{X}) \otimes \mathbb{Z}G$ and $H_2(\tilde{Z})$ are $n(1/p + 1/q - 1/n_0)$ and $n(1/a + 1/b) - 1$ respectively. The result follows.

Corollary 4.2. *The following generalized triangle presentations each admit nonempty efficient spherical pictures:*
(i) $\langle a, b \mid a^2 = b^3 = ((ab)^4 (ab^2)^2)^2 = 1 \rangle$;
(ii) $\langle a, b \mid a^3 = b^3 = (abab^2)^2 = 1 \rangle$.

Proof. The first of these is the group 12.54 of Conder's list [13]. The third relator r^2 is uniquely exceptional: it is a cyclic permutation of $(bUbU^{-1})^2$, where $U = ababa$. The uniqueness stems from the fact that b occurs four times in r and its inverse b^2 occurs only twice. It is thus of type $(3, 3, 2)$. But it has order 2880, so the equation of the theorem fails.

The second group is again uniquely of type $E(3, 3, 2)$, and a coset enumeration shows it to have order 180, so again the equation does not hold.

In both examples the factor groups $A = \langle a \rangle$ and $B = \langle b \rangle$ are readily seen to embed in G, and the root r of the third relator has order 2 in G. It follows immediately from the Theorem that in each case there are nonempty efficient spherical pictures.

Remarks. (1) We can now summarise the status of Conjecture F. Our results show that the conjecture holds for $m \geq 4$, and also for $m = 3$ in the case where r contains no letter of order 2. On the other hand, the conjecture is false in general for $m = 2$ and for $m = 3$, as shown by Figures 7 and 8 respectively. One could propose a weaker form of the conjecture, saying for example that in any efficient picture on D^2 with at least one interior vertex, at least m vertices are connected to D^2. Such a statement would still suffice to prove the Freiheitssatz, and to extend some of the other results of §3. The above corollary shows that no such weaker conjecture can hold when $m = 2$, but what of the case $m = 3$?

(2) The proof above is non-constructive, in the sense that it shows the existence of efficient spherical pictures via an algebraic calculation, but does not actually produce such a picture. We will rectify this by giving a specific example below.

Example.   Consider case (ii) of the Corollary above. In this case G is a direct product $A_5 \times C_3$ (we are grateful to Derek Holt and Rick Thomas for comments about this), while $G_0 \cong A_4$. The kernel of the natural map from G to C_3 is the normal closure of a. It contains G_0, and in particular it contains r. Let Λ denote the group ring $\mathbb{Z}_6 C_3$, and let $\psi : \mathbb{Z}G \to \Lambda$ be the natural map. Given any spherical picture Γ, we can choose basepoints p_v for each vertex v of Γ, such that the label of v read from p_v is precisely r^m. Among the p_v, pick a basepoint p for the whole picture.

The *evaluation* of Γ is defined to be the element $\epsilon(\Gamma) := \sum_v \epsilon_v \lambda_v \in \mathbb{Z}G$, where $\epsilon_v = \pm 1$ is the orientation of v and λ_v is the label of any transverse path in Γ from p to p_v. See Pride [52] for a full discussion. The evaluation does not depend on the choice of *spray* (collection of transverse paths), but is sensitive to the initial choice of basepoints. Change of basepoint at a single vertex adds a multiple of $1 - r$ to the evaluation, while change of overall basepoint (amongst the p_v) multiplies the evaluation by an element of G. Bridge moves do not alter the evaluation, nor does elimination of cancelling pairs of vertices (provided their basepoints are chosen in a compatible way).

Finally, the spherical picture $\Gamma(3,3,2)$ (Figure 6) satisfies $\psi\epsilon(\Gamma(3,3,2)) = 0$. It follows that $\psi\epsilon(\Gamma) = 0$ for any picture Γ that is equivalent to the empty picture via bridge moves, insertion and deletion of cancelling pairs, and insertion and deletion of floating $\Gamma(3,3,2)$'s.

Now consider the picture Γ in Figure 9. It is easy to calculate that $\psi\epsilon(\Gamma) = 3(1 + b)b^t$ for some $t \in \{0,1,2\}$ depending on the choice of base-point. Hence Γ is not equivalent to an empty picture. To see that Γ is efficient, one notes first that, since allowable moves change numbers of vertices by multiples of 2, any smaller picture equivalent to Γ would have an even number of vertices. Secondly, since $\epsilon(\Gamma)$ has augmentation 6, any picture equivalent to Γ has at least 6 vertices. Thirdly, by the Euler formula, and the fact that any vertex in a spherical picture over G has at least 4 neighbours, we can see that a 6-vertex or 8-vertex spherical picture over G would have to contain at least one pair of adjacent vertices, each with exactly 4 neighbours. It is not hard to deduce that such a picture must contain (up to bridge moves) a 4-vertex subpicture of $\Gamma(3,3,2)$, contrary to the hypothesis of efficiency. This leaves us to consider 10-vertex spherical pictures. A slightly longer argument along similar lines (the details of which we omit) shows that no 10-vertex spherical picture can be equivalent to Γ.

Final remark.   All the counterexamples to Conjecture F that we know about satisfy the Freiheitssatz. The evidence remains strong that the Freiheitssatz

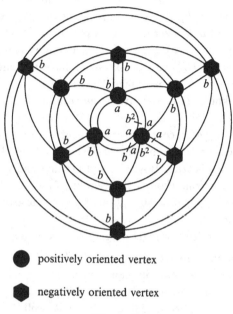

positively oriented vertex

negatively oriented vertex

Figure 9.

holds for any one-relator product in which the relator is a proper power. A proof of this assertion, however, does not yet appear to be in sight.

We have noted in §3 above that in some cases where the relator is exceptional Theorem 3.6 and its corollaries may fail. This is true even in cases where Conjecture F holds, although some analogous statement can then usually be proved, for example Theorem 3.9.

Similarly, we know of no counterexamples to the Cohen Lyndon Theorem (3.13), the Magnus Theorem (3.14), or the solubility of the generalized word and genus problems (3.15, 3.16), for any one-relator product with proper power relator r^m. However, our methods break down when m becomes very small. For example, the Spelling Theorem (3.10) gives insufficient information to prove Theorems 3.13 and 3.14 in the cases $m = 4$ and $m = 5$, and no information whatsoever in the case $m = 2$.

5. Applications

In this section we indicate some applications of the results mentioned in earlier sections, and also some other applications of the method of pictures.

For other applications of pictures, see Pride's survey article [52], and some of the references cited there.

(1) Let k be a knot in S^3, and let M be the manifold formed from S^3 by doing a/b-surgery along k, where a, b are integers with no common prime factor (possibly one of a, b is zero). If M is not prime (that is, $M = M_1 \# M_2$ is a nontrivial connected sum), then $\pi_1(M)$ is a nontrivial free product $A * B$, and the trivial group $\pi_1(S^3)$ can be shown to have the one-relator product $(A * B)/N(r^b)$ as a homomorphic image [29], where $r \in \pi_1(M)$ is the element traced out by the knot. It follows from the results of [29] that $b \leq 5$, and from the Freiheitssatz of §3 that $b \leq 3$. Moreover, $b \leq 2$ unless one of $A = \pi_1(M_1), B = \pi_1(M_2)$ contains an element of order 2 (from which it follows that some connected summand of M has finite fundamental group). In fact, more than this is true. Boyer [6] and Gordon and Luecke [30] showed that $b \leq 1$ in this situation. From a theorem in [15] we can have $b > 1$ only if both M_1 and M_2 are lens spaces (so A, B are finite cyclic groups, and Theorem 3.3 applies). The assertion $b \leq 1$ is best possible, as certain connected sums of lens spaces can arise from integer surgery on knots [30].

(2) Let G be a group, and let $w = w(x) \in G * \langle x \rangle$. Makar-Limanov and Makar-Limanov [48] show that the equation $w(x)^m = 1$ in x can be solved over G (that is, the natural map $G \to (G * \langle x \rangle)/N(w(x)^m)$ is injective) for some positive integer m. Egorov [24] and Taylor-Russell [56] have shown that $w(x)^m = 1$ is solvable over G for all $m \geq 4$ provided either G contains no 2- or 3-torsion [24], or x appears in w with nonzero exponent sum [56]. Our results show that in fact $w^m = 1$ can be solved over G (i) for all $m \geq 4$; (ii) for all $m \geq 3$ if w has no letter of order 2.

(3) Howie and Thomas [40] (see also [41]) studied the two-relator products

$$(2, 3, p; q) = \langle a, b \mid a^2 = b^3 = (ab)^p = [a, b]^q = 1 \rangle$$

of C_2 and C_3, in connection with a conjecture of Coxeter, using the methods described in this article, namely pictures, curvature and cohomology. They showed that if $p \geq 7$ and $q \geq 4$, then with only finitely many exceptions the group $(2, 3, p; q)$ is infinite. The results in the case $p = 7$ were also obtained by Holt and Plesken [32] by other methods. The group $(2, 3, 7; 11)$ was shown to be infinite by Edjvet [21], again using pictures. It was not covered by the results of [32] and [40]. It is now known which of these groups are infinite, except for a single case which remains open - namely $(2, 3, 13; 4)$. The larger class of groups $(l, m, n; p)$, and other related examples, are considered by Chaltin [10].

(4) Finally, we refer the reader to [5, 22, 27] for further applications of pictures to the solution of equations over groups, and related questions.

6. Further problems

We finish by mentioning some open problems suggested by the work described in this article.

(1) Does the Freiheitssatz hold for any one-relator product with relator a proper power? In other words, if $r \in A * B$ is not conjugate to an element of A and $m \geq 2$, is the natural map $A \rightarrow (A * B)/N(r^m)$ injective? We know that the answer is yes in any situation where conjecture F holds, but the Freiheitssatz also holds in all known examples, even in the counterexamples to conjecture F (see §4).

(2) Is conjecture F true when $m = 2$ and r has no letters of order 2 or 3? More generally, is conjecture F true whenever the relator r^m is not exceptional? All the counterexamples given in §4 have exceptional relations containing letters of orders 2 or 3.

(3) Is the weaker from of Conjecture F proposed in Remark 1, §4 true? In other words, given an efficient picture over $(A * B)/N(r^3)$ on D^2 with at most two vertices connected to ∂D^2 by arcs, is it always the case that all vertices are connected to ∂D^2 by arcs?

(4) Is there an analogue of Theorem 3.9 for multiply exceptional relator words? We can show that any given word has at most 3 exceptional forms, and we can completely classify all doubly and triply exceptional words.

(5) Can the work of Anshel [1] and Bogley [4] on 2-relator groups and n-relator groups be generalised to 2-relator products (respectively n-relator products) in some sensible way?

References

[1] I. Anshel, *On two relator groups*, in: Topology and Combinatorial Group Theory (P. Latiolais, ed.), Lecture Notes in Mathematics 1440, Springer, 1990, pp. 1–21.

[2] B. Baumslag, *Free products of locally indicable groups with a single relator*, Bull. Austral. Math. Soc. **29** (1984), pp. 401–404.

[3] G. Baumslag, J. Morgan and P. B. Shalen, *Generalized triangle groups*, Math. Proc. Camb. Phil. Soc. **102** (1987), pp. 25–31.

[4] W. A. Bogley, *An identity theorem for multi-relator groups*, Math. Proc. Camb. Phil. Soc. **109** (1991), pp. 313–321.

[5] W. A. Bogley and S. J. Pride, *Aspherical relative presentations*, Proc. Edinburgh Math. Soc. **35** (1992), pp. 1–39.

[6] S. Boyer, *On proper powers in free products and Dehn surgery*, J. Pure Appl. Alg. **51** (1988), pp. 217–229.

[7]   S. D. Brodskiĭ, *Anomalous products of locally indicable groups*, Algebraicheskie Sistemy, Ivanovo University, 1981, pp. 51–77. (Russian).

[8]   S. D. Brodskiĭ, *Equations over groups and groups with a single defining relation*, Siberian Math. J. **25** (1984), pp. 231–251.

[9]   R. Brown and J. Huebschmann, *Identities among relations*, in: Low Dimensional Topology (R. Brown and T. L. Thickstun, eds.), London Math. Soc. Lecture Notes Series 48, Cambridge University Press, 1982, pp. 153–202.

[10] H. Chaltin, *Among Coxeter's groups $((l, m, n; p))$, 'many' are infinite and do not collapse*, preprint.

[11] D. E. Cohen and R. C. Lyndon, *Free bases for normal subgroups of free groups*, Trans. Amer. Math. Soc. **108** (1963), pp. 528–537.

[12] L. P. Comerford and C. C. Edmunds, *Quadratic equations over free groups and free products*, J. Algebra **68** (1981), pp. 276–297.

[13] M. Conder, *Three-relator quotients of the modular group*, Quart. J. Math. **38** (1987), pp. 427–447.

[14] M. Culler, *Using surfaces to solve equations in free groups*, Topology **20** (1981), pp. 133–145.

[15] M. Culler, C. McA. Gordon, J. Luecke and P. Shalen, *Dehn surgery on knots*, Ann. of Math. **125** (1987), pp. 237–300.

[16] A. J. Duncan and J. Howie, *The genus problem for one-relator products of locally indicable groups*, Math. Z. **208** (1991), pp. 225—237.

[17] A. J. Duncan and J. Howie, *The nonorientable genus problem for one-relator products*, Commun. in Alg. **19** (1991), pp. 2547–2556.

[18] A. J. Duncan and J. Howie, *Weinbaum's conjecture on unique subwords of non-periodic words*, to appear in Proc. Amer. Math. Soc.

[19] A. J. Duncan and J. Howie, *Spelling theorems and Cohen-Lyndon Theorems for one-relator products*, preprint.

[20] M. Edjvet, *A Magnus Theorem for free products of locally indicable groups*, Glasgow Math. J. **31** (1989), pp. 383–387.

[21] M. Edjvet, *An example of an infinite group*, in: Proceedings of a Conference dedicated to A.M. Macbeath (W.A. Harvey, ed.), to appear in LMS Lecture Notes.

[22] M. Edjvet, *On the asphericity of one-relator relative presentations*, preprint.

[23] M. Edjvet and J. Howie, *A Cohen-Lyndon Theorem for free products of locally indicable groups*, J. Pure Appl. Alg. **45** (1987), pp. 41–44.

[24] V. Egorov, *On periodic equations over groups*, Manuscript no. 1127-83, *VINITI* (1983). (Russian).

[25] R. Fenn, *Techniques of Geometric Topology*, London Math. Soc. Lecture Notes 57, Cambridge University Press, 1983.

[26] B. Fine, J. Howie and G. Rosenberger, *One-relator quotients and free products of cyclics*, Proc. Amer. Math. Soc. **102** (1988), pp. 1–6.

[27] S. M. Gersten, *Reducible diagrams and equations over groups*, in: Essays in group theory (S.M. Gersten, ed.), MSRI Publications 8, Springer, 1987, pp. 15–73.

[28] R. Z. Goldstein and E. C. Turner, *Applications of topological graph theory to group theory*, Math. Z. **165** (1979), pp. 1–10.

[29] F. Gonzalez-Acuña and H. Short, *Knot surgery and primeness*, Math. Proc. Camb. Phil. Soc. **99** (1986), pp. 89–102.

[30] C. McA. Gordon and J. Luecke, *Only integral Dehn surgeries can yield reducible manifolds*, Math. Proc. Camb. Phil. Soc. **102** (1987), pp. 97–101.

[31] G. A. Gurevich, *On the conjugacy problem for groups with a single defining relation*, Soviet Math. Dokl. **13** (1972), pp. 1436–1439.

[32] D. Holt and W. Plesken, *A cohomological criterion for a finitely presented group to be infinite*, to appear in J. London Math. Soc.

[33] J. Howie, *On pairs of 2-complexes and systems of equations over groups*, J. reine angew. Math. **324** (1981), pp. 165–174.

[34] J. Howie, *On locally indicable groups*, Math. Z. **180** (1982), pp. 445–461.

[35] J. Howie, *Cohomology of one-relator products of locally indicable groups*, J. London Math. Soc. **30** (1984), pp. 419–430.

[36] J. Howie, *The quotient of a free product of groups by a single high-powered relator. I. Pictures. Fifth and higher powers*, Proc. London Math. Soc. **59** (1989), pp. 507–540.

[37] J. Howie, *The quotient of a free product of groups by a single high-powered relator. II. Fourth powers*, Proc. London Math. Soc. **61** (1990), pp. 33–62.

[38] J. Howie, *The quotient of a free product of groups by a single high-powered relator. III. The word problem*, Proc. London Math. Soc. **62** (1991), pp. 590–606.

[39] J. Howie and S. J. Pride, *A spelling theorem for staggered generalized 2-complexes, with applications*, Invent. Math. **76** (1984), pp. 55–74.

[40] J. Howie and R. M. Thomas, *The groups $(2,3,p;q)$; asphericity and a conjecture of Coxeter*, to appear in J. Algebra.

[41] J. Howie and R. M. Thomas, *Proving certain groups infinite*, these proceedings.

[42] J. Huebschmann, *Cohomology of aspherical groups and of small cancellation groups*, J. Pure Appl. Alg. **14** (1979), pp. 137–143.

[43] R. C. Lyndon, *Cohomology theory of groups with a single defining relation*, Ann. of Math. **52** (1950), pp. 650–665.

[44] R. C. Lyndon and P. E. Schupp, *Combinatorial Group Theory*, Springer, 1977.

[45] W. Magnus, *Über diskontinuierliche Gruppen mit einer definierenden Relation (Der Freiheitssatz)*, J. reine angew. Math. **163** (1930), pp. 141–165.

[46] W. Magnus, *Untersuchungen über unendliche diskontinuierliche Gruppen*, Math. Ann. **105** (1931), pp. 52–74.

[47] W. Magnus, *Das Identitätsproblem für Gruppen mit einer definierenden Relation*, Math. Ann. **106** (1932), pp. 295–307.

[48] L. Makar-Limanov and O. Makar-Limanov, *On equations over groups*, J. Algebra **93** (1985), pp. 165–168.

[49] V.F. Mazurovskiǐ, *On the word problem for anomalous products of groups*, Algebraicheskie Sistemy, Ivanovo University, 1991, pp. 26–35. (Russian).

[50] B. B. Newman, *Some results on one-relator groups*, Bull. Amer. Math. Soc. **74** (1968), pp. 568–571.

[51] S. J. Pride, *Star-complexes and the dependence problem for hyperbolic complexes*, Glasgow Math. J. **30** (1988), pp. 155–170.

[52] S.J. Pride, *Identities among relations of group presentations*, in: Group Theory from a Geometric Viewpoint (E. Ghys, A. Haefliger and A. Verjovsky, eds.), Conference Proceedings, ICTP, World Scientific, 1991, pp. 687—717.

[53] C. P. Rourke, *Presentations and the trivial group*, in: Topology of Low Dimensional Manifolds (R. Fenn, ed.), Lecture Notes in Mathematics 722, Springer, 1979, pp. 134–143.

[54] H. Short, *Topological methods in group theory: the adjunction problem*, Ph. D. thesis, University of Warwick, 1984.

[55] H. Short, *The genus problem for hyperbolic groups*, preprint.

[56] G. Taylor-Russell, *The solution of equations of the form $R^n = 1$ over groups, with generalization to one-relator products*, Ph. D. thesis, University of London, 1990.

[57] C. M. Weinbaum, *On relators and diagrams for groups with a single defining relator*, Illinois J. Math. **16** (1972), pp. 308–322.

[58] M. Wicks, *Commutators in free products*, J. London Math. Soc. **37** (1962), pp. 433–444.

An Inaccessible Group

Martin J. Dunwoody

Faculty of Mathematical Studies, University of Southampton, Highfield,
Southampton SO9 5NH.

1. Introduction

Stallings [6] showed that a group G has more than one end if and only if
$G \approx A *_F B$, where F is finite, $A \neq F \neq B$, or G is an HNN-extension with
finite edge group F.

A finitely generated group G is said to be *accessible* if it is the fundamental
group of a graph of groups in which all edge groups are finite and every vertex
group has at most one end. We say that G is *inaccessible* if it is not accessible.

Let $d(G)$ denote the minimal number of generators of the finitely generated
group G. It follows from Grushko's Theorem that $d(G * H) = d(G) + d(H)$. It
follows that G is a free product of indecomposable groups, i.e. groups which
cannot be written as a non-trivial free product. The problem of accessibility is whether we can replace the free product with free product with finite
amalgamation in the last statement. (The number of HNN-decompositions
is bounded by $d(G)$.) However, there is no analogue of Grushko's Theorem.
In fact, if G is accessible then any process of sucessively decomposing G, and
the factors that arise in the process, terminates after a finite number of steps.
See [2] for a proof of this and related results.

Linnell [5] proved that if G is finitely generated then, for any reduced de-
composition of G as a graph of groups X in which all edge groups are finite,
there is a bound B such that $\sum_{e \in E} 1/|G_e| < B$, where E is the edge set of
X. Thus for any $k > 0$, there are at most kB edges e such that $|G_e| \leq k$. In
[3] I showed that G is accessible if G is almost finitely presented. Groves and
Swarup [4] have extended this result to a somewhat larger class of groups.
This paper contains the construction of a finitely generated inaccessible group.
C.T.C. Wall [8] conjectured that all finitely generated groups are accessible.
On the other hand, Bestvina and Feighn [1] have given an example of a finitely
generated group which does not satisfy a generalized accessibility condition

in which decompositions over torus subgroups are allowed. It was by thinking about their construction that I thought of the example presented here.

Let X be a connected locally finite graph. Thomassen and Woess [7] have defined X to be accessible if for some positive integer n any pair of ends of X can be separated by removing at most n edges. They show, by using results from [2] Chapter 2, that a finitely generated group G is accessible as a group if and only if its Cayley graph (with respect to a finite generating set) is accessible as a graph. They investigate alternative definitions for a graph to be accessible.

I am very grateful to Warren Dicks, Peter Kropholler and Martin Roller for providing short proofs that the group J is inaccessible to replace my laboured argument.

2. Constructing the example

Suppose we have a lattice of groups as shown.

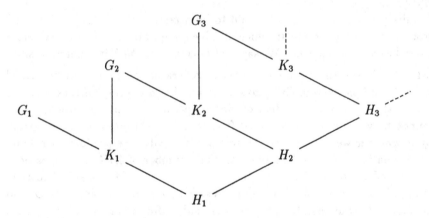

In the diagram lines represent proper inclusions. We also require that G_{i+1} is generated by K_i and H_{i+1}.

We show how to associate an inaccessible group with such a group lattice, when K_i (and hence H_i) is finite for all i, and G_1 is finitely generated. In the next section we show that such a lattice of groups exists.

Let P be the fundamental group of the graph of groups

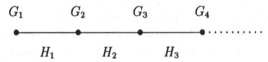

In P, we have $H_1 < H_2 < \ldots < H_\omega := \bigcup_{i \in \mathbb{N}} H_i$. Since H_ω is countable it can be embedded in a finitely generated group H. Let J be the free product with amalgamation $P *_{H_\omega} H$. Now J is generated by G_1 and H. For suppose L is the subgroup generated by G_1 and H. It suffices to show that $G_i < L$ for all $i \in \mathbb{N}$. But if $G_i < L$, then $K_i < L$ and $H_{i+1} < L$ and so $G_{i+1} < L$. It follows by induction that $G_i < L$ for all $i \in \mathbb{N}$. Hence J is finitely generated.

Let P_n be the fundamental group of the graph of groups

and let Q_n be the fundamental group of the graph of groups

Thus $P = P_n *_{K_n} Q_n$. Since $H_\omega < Q_n$, J decomposes as

$$J = P_n \underset{K_n}{*} (Q_n \underset{H_\omega}{*} H),$$

and so if $J_n = Q_n *_{H_\omega} H$, J decomposes as the fundamental group of the graph of groups

It follows immediately that J is inaccessible.

3. Constructing the lattice

In this section we construct a lattice of groups as specified in the previous section. Let H be the subgroup of $\mathrm{Symm}(\mathbb{Z})$ generated by the transposition $t = (0,1)$ and the shift map s, where $s(i) = i + 1$. Put $t_i = s^i t s^{-i} = (i, i+1)$. Let $H_i = \langle t_{-i}, t_{-i+1}, \ldots, t_0, t_1, \ldots, t_{i-1} \rangle$. Thus H_i is isomorphic to the symmetric group S_{2i+1}. Let V be the group of all maps $\mathbb{Z} \to \mathbb{Z}_2$ with finite support, under the usual addition. Then H acts on V by $vh(n) = v(h(n))$ for all $v \in V$, $h \in H$ and $n \in \mathbb{Z}$. Let V_i be the subgroup of V consisting of all maps with support $[-i,i] = \{-i, -i+1, \ldots, 0, 1, \ldots, i\}$. Let $G_i' = V_i \rtimes H_i$. Let $z_i \in V_i$ with $z_i(n) = 1$ for $n \in [-i,i]$, then z_i is central in G_i'. Let $K_i = \langle z_i, H_i \rangle$ and note that $K_i = \mathbb{Z}_2 \times H_i$. For $i = 1, 2, \ldots$ let G_i be an isomorphic copy of G_i' and identify K_i with its image in G_i. We can then assume that $K_i = G_i \cap G_{i+1}$. It is left to the reader to check that the lattice of groups is as required.

References

[1] M. Bestvina and M. Feighn, *A counterexample to generalized accessibility*, in: Arboreal Group Theory, MSRI Publications 19, Springer, 1991, pp. 133–142.

[2] W. Dicks and M. J. Dunwoody, *Groups Acting On Graphs*, Cambridge University Press, 1989.

[3] M.J. Dunwoody, *The accessibility of finitely presented groups*, Invent. Math. **81** (1985), pp. 449–457.

[4] J.R.J. Groves and G.A. Swarup, *Remarks on a technique of Dunwoody*, J. Pure Appl. Algebra **75** (1991), pp. 259–269.

[5] P. A. Linnell, *On accessibility of groups*, J. Pure Appl. Algebra **30** (1983), pp. 39–46.

[6] J. R. Stallings, *Group Theory And Three-Dimensional Manifolds*, Yale Math. Monographs 4, Yale University Press, 1971.

[7] C. Thomassen and W. Woess, *Vertex-transitive graphs and accessibility*, preprint, The Technical University and Universita di Milano, 1991, to appear in J. Combin. Th. (Ser. B).

[8] C. T. C. Wall, *Pairs of relative cohomological dimension one*, J. Pure Appl. Algebra **1** (1971), pp. 141–154.

Isoperimetric and Isodiametric Functions of Finite Presentations

Steve M. Gersten

Mathematics Department, University of Utah, Salt Lake City, UT 84112, USA.

1. Introduction and definitions

Isoperimetric functions are classical in differential geometry, but their use in group theory derives from Gromov's seminal article [Gr] and his characterization of word hyperbolic groups by a linear isoperimetric inequality. Isodiametric functions were introduced in our article [G1] in an attempt to provide a group theoretic framework for a result of Casson's (see Theorem 3.6 below). It turned out subsequently that the notion had been considered earlier under a different name [FHL]. We have learned since that the differential geometers also have their isodiametric functions and they mean something different by them. However the analogy is too suggestive to abandon this terminology and we shall retain it here. Up to an appropriate equivalence relation (Proposition 1.1 below), isoperimetric and isodiametric functions are quasiisometry invariants of finitely presented groups. Hence these functions are examples of *geometric properties*, in the terminology of [Gh].

If $\mathcal{P} = \langle x_1, x_2, \ldots, x_p \mid R_1, R_2, \ldots R_q \rangle$ is a finite presentation, we shall denote by $G = G(\mathcal{P})$ the associated group; here $G = F/N$, where F is the free group freely generated by the generators $x_1, \ldots, x_p$ and N is the normal closure of the relators. If w is an element of F (which we may identify with a reduced word in the free basis), we write $\ell(w)$ for the length of the word w and $\bar{w}$ for the element of G represented by w. We shall use freely the terminology of van Kampen diagrams [LS, p. 235ff] in the sequel.

We write $\text{Area}_\mathcal{P}(w)$ for the minimum number of faces (i.e. 2-cells) in a van Kampen diagram with boundary label w. Equivalently, $\text{Area}_\mathcal{P}(w)$ is the minimum number of relators or inverses of relators occurring in all expressions of w as a product (in F) of their conjugates. The function $f : \mathbb{N} \to \mathbb{N}$ is an

This research was partially supported by the NSF. In addition it is a pleasure to thank the University of Sussex for their kind hospitality.

isoperimetric function for $\mathcal{P}$ if, for all n and all words w with $\ell(w) \leq n$ and $\bar{w} = 1$, we have $\mathrm{Area}_{\mathcal{P}}(w) \leq f(n)$. The minimum such isoperimetric function is called the *Dehn function* of $\mathcal{P}$.

If $\mathcal{D}$ is a van Kampen diagram with boundary label w, we choose the base point v_0 in the boundary of $\mathcal{D}$ corresponding to where one starts reading the boundary label w and one defines

$$\mathrm{Diam}_{v_0}(\mathcal{D}) = \max_{v \in \mathcal{D}^{(0)}} d_{\mathcal{D}^{(1)}}(v_0, v).$$

Here $d_{\mathcal{D}^{(1)}}$ denotes the word metric for the 1-skeleton of $\mathcal{D}$, so that every edge has length 1. The function $f : \mathbb{N} \to \mathbb{N}$ is called an *isodiametric function* for $\mathcal{P}$ if, for all n and all reduced words w with $\ell(w) \leq n$ and $\bar{w} = 1$, there exists a based van Kampen diagram $(\mathcal{D}, v_0)$ for w with $\mathrm{Diam}_{v_0}(\mathcal{D}) \leq f(n)$. A more algebraic way of formulating this is as follows. Let M denote the maximum length of a relator of $\mathcal{P}$. Let f be an isodiametric function for $\mathcal{P}$. If $\bar{w} = 1$, then one can write

$$w = \prod_{i=1}^{m} R_{j_i}^{\epsilon_i u_i},$$

where R_{j_i} is a relator, $\epsilon_i = \pm 1$, $u_i \in F$ and $\ell(u_i) \leq f(\ell(w)) + M$. Here we write $a^b = bab^{-1}$ for elements a and b in a group.

A word of caution is necessary here. A diagram of minimal area is always reduced, in the sense of [LS]. However this will not be the case in general for a diagram of minimal diameter. This complicates considerably the problem of proving that a diagram is diametrically minimal. Consequently we do not introduce a diametric analog of the Dehn function.

Next we discuss the question of change of presentation.

Proposition 1.1. [Al], [Sh] *Let $\mathcal{P}$ and $\mathcal{P}'$ be finite presentations for isomorphic groups. If f is an isoperimetric function (resp. isodiametric function) for $\mathcal{P}$, then there exist positive constants $A, B, C, D,$ and E such that $n \mapsto Af(Bn + C) + Dn + E$ is an isoperimetric (resp. isodiametric) function for $\mathcal{P}'$.*

In fact, isoperimetric (resp. isodiametric) functions transform in the same way for quasiisometric presentations, so, up to the obvious equivalence relation, these are quasiisometry invariants (for the notion of quasiisometry, see [Gh]). In particular it makes sense to speak of a finitely presented group possessing a linear, quadratic, polynomial, exponential, etc., isoperimetric (resp. isodiametric) function, or more loosely, to speak of the group satisfying the the appropriate isoperimetric (resp. isodiametric) inequality.

Example. If one takes $\mathcal{P}$ to be a presentation with no relators, then the area function is identically zero, so the Dehn function is zero. However, a presentation of a free group with defining relators will have a nonzero Dehn function. Thus the awkward constants D and E in Proposition 1 are in general necessary.

Remark. An interesting variation on the notion of isoperimetric function was suggested by Gromov. We consider words w which are boundary labels of diagrams whose domains are compact orientable surfaces of some (variable) genus. Equivalently, we may assume there are words $u_1, u_2, \ldots, u_g, v_1, v_2, \ldots, v_g$ such that $w' = w \prod_{i=1}^{g} [u_i, v_i]$ represents 1 in the group G of our finite presentation $\mathcal{P}$; here $[u_i, v_i]$ denotes the formal commutator $u_i v_i u_i^{-1} v_i^{-1}$, where u_i^{-1} denotes the formal inverse of the word u_i (invert each letter and write them in the reverse order); write $w \sim 0$ if there exists $g \geq 0$ such that this condition is satisfied.

We define $\text{Area}'_{\mathcal{P}}(w) = \min_{w'} \text{Area}_{\mathcal{P}}(w')$, where the minimum is taken over all words w' constructed from w in this way. Then we define

$$f'(n) = \max_{\substack{w \sim 0 \\ \ell(w) \leq n}} \text{Area}'_{\mathcal{P}}(w).$$

Gromov remarks that the function f' is closer in spirit to the differential geometric notion of minimal surface spanned by a loop, where one cannot control the genus of the (orientable) surface spanned.

2. Relation with the word problem

The functions introduced in §1 are important for discussing the complexity of the word problem for a finitely presented group.

Theorem 2.1. *The following are equivalent for a finite presentation $\mathcal{P}$.*

2.1.1. $G(\mathcal{P})$ has a solvable word problem.

2.1.2. $\mathcal{P}$ has a recursive isoperimetric function (in which case, the Dehn function itself is recursive).

2.1.3. $\mathcal{P}$ has a recursive isodiametric function.

Let us sketch the argument. For the implication (1)$\Rightarrow$(2), we solve the word problem for all words of length at most n, thereby obtaining for each word w satisfying $\ell(w) \leq n$ and $\bar{w} = 1$ in G some expression

$$w = \prod_{i=1}^{k(w)} R_{j_i}^{\epsilon_i u_i}$$

in the free group F. Let us define a function f by

$$f(n) = \sup_{\substack{\ell(w) \leq n \\ \bar{w}=1}} k(w).$$

Then f is a recursive isoperimetric function for $\mathcal{P}$.

The implication (2)$\Rightarrow$(3) follows from the following elementary result.

Lemma 2.2. *If f is an isoperimetric function for $\mathcal{P}$, then $n \mapsto Mf(n)+n$ is an isodiametric function for $\mathcal{P}$, where M is the maximum length of a relator.*

Proof. Let w be a word of length n representing 1 in $G = G(\mathcal{P})$ and let $\mathcal{D}$ be a van Kampen diagram of minimal area for w. If V and F denote the number of vertices and faces of $\mathcal{D}$, we observe that the length of the longest edge path in $\mathcal{D}^{(1)}$ which does not contain a circuit is at most $V-1 \leq MF+n$. From this it follows that $\mathrm{Diam}_{v_0}(\mathcal{D}) \leq Mf(n) + n$, and $n \mapsto Mf(n) + n$ is an isodiametric function for $\mathcal{P}$.

The implication (3)$\Rightarrow$(1) proceeds as follows. Let f be an isodiametric function for $\mathcal{P}$. Suppose $\ell(w) = n$ and $\bar{w} = 1$, with $G = G(\mathcal{P})$. Then there is a description

$$w = \prod_{i=1}^{m} R_{j_i}^{\epsilon_i u_i},$$

with $\ell(u_i) \leq f(n) + M$, with M as above. But the set

$$S_m = \{ R^u \mid R \text{ a relator}, \ \ell(u) \leq m \}$$

is finite and generates a finitely generated subgroup $N_m < N < F$. Since the problem of deciding whether or not a word lies in a given finitely generated subgroup of the free group F is effectively solvable, we first calculate $m = f(n) + M$ and then apply this algorithm to decide whether or not $w \in N_m$. This solves the word problem for G.

A particularly attractive geometric way of deciding whether or not $w \in N_m$ as above has been given by Stallings [St1]. His algorithm amounts to using an immersion of finite graphs as a finite state automaton. Note that the automaton depends on the word w being tested.

Remark. It is somewhat mysterious that one has to proceed from Theorem 2.1.3 to 2.1.2 via 2.1.1, thereby involving the complications of general recursive functions. A more satisfying situation is to have a formula for an isoperimetric function in terms of an isodiametric function. One conjecture, which does not contradict any known example, is that there should be an isoperimetric function of the form $n \mapsto a^{f(n)+n}$, for a constant a (Stallings raised this question in the special case when f is linear). In this connection, D. E. Cohen has recently shown [C] that *if f is an isodiametric function for a finite presentation $\mathcal{P}$, then there are positive constants a, b so that $n \mapsto a^{bf(n)+n}$ is an isoperimetric function for $\mathcal{P}$.* His proof makes use of an analysis of Nielsen's reduction process for producing a basis for a subgroup of a free group (see also [G4] for a different treatment involving Stallings' folds).

Here is a striking example, which shows that the complexity of the word problem for 1-relator groups, as measured by the growth of an isodiametric function, can be quite large.

Example. Each isodiametric and each isoperimetric function for the presentation $\mathcal{P} = \langle x, y \mid x^{x^y} = x^2 \rangle$ grows faster than every iterated exponential [G1].

Remark. Magnus showed that all 1-relator groups have a solvable word problem [LS]. However his argument gives no indication of the complexity of the algorithm. It is of interest to determine how fast the Dehn function of a 1-relator presentation can grow. We have shown (unpublished) that Ackermann's function f_ω is (up to the equivalence relation of Proposition 1.1) an isoperimetric function for every 1-relator presentation. Here one defines functions $f_\alpha : \mathbb{N} \to \mathbb{N}$ for ordinals $\alpha \le \omega$ (where ω is the first infinite ordinal) inductively by $f_1(s) = 2s$, $f_{n+1}(s) = f_n^{(s)}(s)$, where $f_n^{(s)}$ denotes the s-fold iterate of f_n, and $f_\omega(s) = f_s(s)$.

The central tool in proving these upper bounds for the Dehn functions of 1-relator presentations is the rewrite function for a pair (G, H), where H is a finitely generated subgroup of the finitely generated group G. We suppose A, B are finite sets of generators for G, H, respectively, and we let $|g|_{G,A}$ denote the distance of $g \in G$ from the identity in the word metric, and similarly define $|h|_{H,B}$ for $h \in H$. We let

$$f_{G,H}(n) = \max_{\substack{h \in H \\ |h|_{G,A} \le n}} |h|_{H,B},$$

and we call $f_{G,H}$ the *rewrite function*. We calculate $f_{G,H}$ inductively for a 1-relator presentation and Magnus subgroup and then apply the result to calculate an isoperimetric function. The rewrite function bears the same relation to the generalized word problem that the Dehn function bears to the word problem.

It is an open question whether for each n one can find a 1-relator presentation whose Dehn function grows at least as fast as f_n. The Dehn function for the presentation $x^{x^y} = x^2$ grows at least as fast as f_3, but this is the fastest growth we have actually proved can be realized for 1-relator presentations [G2].

3. Examples and applications

By Proposition 1.1, the simplest invariant condition on isoperimetric functions is that they be linear. In this case there is a satisfactory characterization. If we have a given finite set of generators A for a finitely presented group G, then for sufficiently large N, the presentation $\mathcal{P}_N$, with generators A and relators consisting of all relations among the generators of length at most N, will be a presentation of G.

Theorem 3.1. *The following are equivalent for a finitely presented group G.*

3.1.1. G has a linear isoperimetric function.

3.1.2. G is word hyperbolic.

3.1.3. There exists a finite presentation for G which satisfies Dehn's algorithm.

3.1.4. If A is a finite set of generators for G, then for all sufficiently large N, the presentation $\mathcal{P}_N$ for G satisfies Dehn's algorithm.

The unexplained terms in the theorem are as follows. Let A be a finite set of generators for G and let Γ be the associated Cayley graph, equipped with the word metric. The group G is called word hyperbolic if Γ is δ-hyperbolic for some $\delta \geq 0$; here Γ is called δ-hyperbolic if every geodesic triangle Δ in it satisfies Rips's condition R_δ: every point on one side of Δ is at distance at most δ from the union of the other two sides. The finite presentation $\mathcal{P}$ for G is said to satisfy Dehn's algorithm if, given any nonempty word w with $\bar{w} = 1$, there is a relator R of $\mathcal{P}$ such that w contains greater than $\frac{1}{2}$ of the word R as a contiguous subword.

The proof of Theorem 3.1 is very attractively presented in [ABC].

The next step beyond linear is subquadratic isoperimetric functions. In this case, Gromov asserts that a finite presentation with a subquadratic isoperimetric function also possesses a linear isoperimetric function [Gr, 2.3.F]. A. Yu. Ol'shanskii recently found an elementary proof of this important result [Ol].

In order to explain how quadratic isoperimetric functions arise, it is necessary to introduce new notions.

Definition 3.2. *Let G be a finitely generated group with finite set A of semigroup generators and associated Cayley graph Γ. One has the evaluation mapping $A^* \to G$, $w \mapsto \bar{w}$, where A^* is the free monoid on A. Such a word w can be viewed as a path $w(t)$, $t \geq 0$, parametrized by arc length for $t \leq \ell(w)$, starting at the base point 1 (where G is identified equivariantly with the vertex set of Γ), moving over an edge in unit time, until it reaches its end point $\bar{w}$ at time $\ell(w)$; from then on, $w(t)$ remains constant at the vertex $\bar{w}$. A combing is a section $\sigma : G \to A^*$ of the evaluation mapping such that there exists a constant $k > 0$ such that*

$$(3.2.1) \quad \forall g \in G \ \forall a \in A \ \forall t \geq 0 \qquad \text{one has} \qquad |\sigma(ga)(t) - \sigma(g)(t)| \leq k;$$

here $|x - y|$ denotes the distance from x to y in Γ. The condition (3.2.1) is called the k-fellow traveller condition. The finitely generated group G is called combable if it admits a combing. In addition we say that σ is linearly bounded if there are constants $C, D > 0$ such that $\ell(\sigma(g)) \leq C|g| + D$ for all $g \in G$, where $|g| := |1 - g|$.

The combing σ is called an automatic structure if the subset $\sigma(G) \subset A^\star$ is a regular language; that is, $\sigma(G)$ is the precise language recognized by a finite state automaton. It is a result of [ECHLPT] *that an automatic structure σ is linearly bounded.*

Remark. The definition of combing adopted in [ECHLPT] is more restrictive than the one we have adopted, following [Gh, p. 26], [Sh]: the former definition implies linear boundedness. It is known that both combability and the existence of a linearly bounded combing are quasiisometry invariant conditions [Sh]. It is unknown whether the existence of an automatic structure is quasiisometry invariant, although one of the results of [ECHLPT] is an algorithm which enables one to translate an automatic structure from one finite set of semigroup generators to another.

Theorem 3.3. *If G is a finitely generated group with a linearly bounded combing, then G is finitely presented and admits a quadratic isoperimetric function.*

We give the proof, which is due to Thurston, of this important result; the reader may consult [ECHLPT], where the result is proved in conjunction with higher dimensional isoperimetric inequalities.

Suppose that A is a finite set of semigroup generators for G and that $\sigma : G \to A^\star$ is a linearly bounded combing. Let σ satisfy the k-fellow traveller condition. Let $w \in A^\star$ be such that $\bar{w} = 1$, so w represents a closed path based at the vertex 1 in the Cayley graph Γ and let $\ell(w) = n$. We shall construct a finite presentation $\mathcal{P}$ for G and a van Kampen diagram for w in $\mathcal{P}$. Let $\sigma_i = \sigma(w(i))$ for integers $0 \leq i \leq n$, so σ_i is a path in Γ from 1 to $w(i)$. Observe that

$$|\sigma_i(j) - \sigma_{i+1}(j)| \leq k,$$

for each integral time j. This means that we can consider the vertices $\sigma_i(j)$, $\sigma_i(j+1)$, $\sigma_{i+1}(j+1)$, and $\sigma_{i+1}(j)$ as lying on a quadrilateral Q_{ij} whose boundary label is a relation of length at most $2k + 2$. If we take the presentation $\mathcal{P}$ to consist of A as generators and as relators, all relations among these generators of length at most $2k + 2$, then we see that w is a consequence of these relators. Consequently $\mathcal{P}$ is a finite presentation for G. Observe that we have only used the k-fellow traveller property so far and not the linear boundedness of the combing.

The quadrilaterals Q_{ij} fit together to form a van Kampen diagram $\mathcal{D}$ for w. Observe that for fixed i we can cut off j at time $\max(\ell(\sigma_i), \ell(\sigma_{i+1}))$, since both paths, σ_i and σ_{i+1}, will have reached their end points by then. But $\ell(\sigma_i) \leq Ci + D$, since $|w(i)| \leq i$. It follows that the total number of quadrilaterals in $\mathcal{D}$ is at most

$$\sum_{i=0}^{n}(Ci + D) \leq An^2 + B$$

for constants $A, B > 0$. Thus $\text{Area}_P(w) \leq A\ell(w)^2 + B$, and the theorem is established.

Remark. Thurston asserts that the $(2n+1)$-dimensional integral Heisenberg group for $n \geq 2$ and $SL_n(\mathbb{Z})$, for $n \geq 4$, satisfy the quadratic isoperimetric inequality [ECHLPT]; no details are available at this time. It would appear then that there was no simple characterization of groups satisfying the quadratic isoperimetric inequality. It is proved in [ECHLPT] and [G2] that the 3-dimensional integral Heisenberg group satisfies a cubic isoperimetric inequality (see also Section 5 below). Furthermore, it is shown in [ECHLPT] that an isoperimetric function for $SL_3(\mathbb{Z})$ must grow at least exponentially. Compare also the arguments sketched in [Gr2].

Proposition 3.4. *If the group G is combable, then it satisfies the linear isodiametric inequality.*

Proof. Let $\sigma : G \to A^*$ be a combing, where A is a finite set of semigroup generators. Suppose σ satisfies the k-fellow traveller property. As in the first part of the proof of Theorem 3.3, we obtain a finite presentation for G whose relators are all words in $w \in A^*$ satisfying $\bar{w} = 1$ and such that $\ell(w) \leq 2k+2$. Furthermore, we obtain a van Kampen diagram $\mathcal{D}$ for w, as in the second part of the proof, except now we have no bounds on the lengths of the paths σ_i. Nevertheless, if we consider a vertex $\sigma_i(j)$, then by holding j fixed and letting i vary, we arrive at the boundary of $\mathcal{D}$ in at most $k\ell(w)$ steps; once we arrive at the boundary, then we can follow it in at most $\ell(w)$ additional steps to arrive at the base point. Since no vertex of $\mathcal{D}$ is farther than k from a vertex of type $\sigma_i(j)$, it follows that the distance in the word metric of $\mathcal{D}^{(1)}$ from the base point to any vertex is bounded by $(k+1)\ell(w) + k$. This establishes the linear isodiametric inequality.

Remark. It is asserted in [Gh, p. 27] that $SL_3(\mathbb{Z})$ is not combable, where the definition of combability adopted there is the same as ours. Thurston's result (Theorem 3.3) is quoted for the proof. However, this last result applies only for a linearly bounded combing, so it must be considered open whether $SL_3(\mathbb{Z})$ is combable or not. We proved in [G3] that all combable groups satisfy an exponential isoperimetric inequality, and this is the best result known to date in this generality.

Remark. There is an analogous notion of *asynchronously combable group*, where one has a section $\sigma : G \to A^*$ satisfying the *k-asynchronous fellow traveller property*: after a monotone reparametrization, the paths $\sigma(g)$ and $\sigma(ga)$ are k-fellow travellers, for $g \in G$ and $a \in A$. The asynchronous combing σ is called an *asynchronously automatic structure* if the language $\sigma(G) \subset A^*$ is regular. M. Shapiro has recently proved that the definition just given is equivalent to that of [ECHLPT] for an asynchronously automatic structure on a group [Sp2]. One can show that *every asynchronously combable group is finitely presented and satisfies a linear isodiametric inequality*. Furthermore, *if G is asynchronously automatic, then it has an exponential isoperimetric function*[ECHLPT], [BGSS].

Remark. In the original version of this survey, we raised the question whether the integral Heisenberg group was combable. Gromov asserted at the conference that the real Heisenberg group is combable. There are several arguments sketched in [Gr2], but our attempts to fill in the details have only succeeded in proving the weaker result that the group is asynchronously combable; so we regard this as an open question of great interest. In this connection, we mention a recent result of M. Bridson's [Bd], that the group $\mathbb{Z}^n \rtimes_\phi \mathbb{Z}$ is asynchronously combable for all $\phi \in \mathrm{Gl}_n(\mathbb{Z})$.

Theorem 3.5. *The following finitely presented groups all have linear isodiametric functions.*

3.5.1. Lattices in the 3-dimensional Lie group Nil.

3.5.2. Lattices in the 3-dimensional Lie group Sol.

3.5.3. $\pi_1(M)$, where M is a compact 3-manifolds for which Thurston's geometrization conjecture [Th] holds.

The statements about lattices in Nil and Sol are proved in [G1]. Here is an extremely rough sketch for lattices in Sol. The problem is reduced to showing that the "Fibonacci group" $\mathbb{Z}^2 \rtimes_\phi \mathbb{Z}$, where $\phi = \begin{pmatrix} 1 & 1 \\ 1 & 0 \end{pmatrix}$, satifies a linear isodiametric inequality. This is deduced from arithmetic properties of the Fibonacci sequence.

The argument for 3.5.3 is as follows. A result of [ECHLPT] states that if no geometric piece in the Thurston decomposition is a Nil or Sol group, then the fundamental group is automatic. Since the Nil and Sol pieces in the Thurston decomposition occur only as connected summands, it follows that the fundamental group $\pi_1(M)$ of a compact 3-manifold M for which Thurston's geometrization conjecture holds is the free product of an automatic group with a finite free product of Nil and Sol groups. Since each of these free factors satisfies the linear isodiametric inequality and since the class of finitely presented groups satisfying the linear isodiametric inequality is closed under finite free products, it follows that $\pi_1(M)$ satisfies the linear isodiametric inequality.

Remark. Bridson's recent results [Bd] strengthen Theorem 3.5, showing that the fundamental group of every compact 3-manifold satisfying Thurston's geometrization conjecture is asynchronously combable. Bridson raises the question of the "logical complexity" of the language of the combing.

Remark. It is not known how wide the class of finitely presented groups satisfying a linear isodiametric inequality is. For instance, we do not know an example of a finitely presented linear group which does not satisfy a linear isodiametric inequality (the example $x^{x^y} = x^2$ given in §2 is not a linear group). Since a finitely presented linear group has a solvable word problem, Theorem 2.1 will be of no help in constructing an example.

Remark. It follows from results of [ECHLPT] that if a compact 3-manifold satisfies Thurston's geometrization conjecture, then its fundamental group has an exponential isoperimetric function. If, in addition, there are no Nil or Sol pieces, it is automatic and satisfies the quadratic isoperimetric inequality.

Question. If ϕ is an automorphism of the finitely generated free group F and if $G = F \rtimes_\phi \mathbb{Z}$ is the corresponding split extension, does G satisfy the quadratic isoperimetric inequality? The result of [BF], that G is word hyperbolic if and only if it contains no subgroup isomorphic to $\mathbb{Z}^2$, can be viewed as positive evidence. Furthermore, G is automatic if ϕ is geometric (that is, if ϕ is induced by a homeomorphism of a compact surface with nonempty boundary). For in this case, G is $\pi_1(M)$, where M is a compact Haken 3-manifold, and Thurston's geometrization conjecture is known to hold for such M [Th]. A cohomological dimension argument shows that M has no Sol or Nil pieces, whence, by the preceding Remark, G is automatic.

That this question may be delicate is suggested by our result (unpublished) that if $\phi \in \text{Aut}(F(a,b,c))$ is given by $\phi(a) = a$, $\phi(b) = ba$, $\phi(c) = ca^2$, then $G = F(a,b,c) \rtimes_\phi \mathbb{Z}$ cannot act properly discontinuously and cocompactly on any geodesic metric space satisfying Gromov's condition CAT(0) (see [GH] for the CAT(0) property).

The original motivation for introducing isodiametric functions was a result proved by Casson in 1990. We shall state a weaker version of his result which falls naturally within our framework. We say that the finite presentation $\mathcal{P}$ satisfies condition ID(α), where $\alpha > 0$, if there is an $\epsilon \geq 0$ so that $n \mapsto \alpha n + \epsilon$ is an isodiametric function for $\mathcal{P}$. This is of course just a reformulation of a linear isodiametric inequality.

Theorem 3.6. [SG] *Let M be a closed, orientable, irreducible, aspherical 3-manifold whose fundamental group admits a finite presentation satisfying condition* ID(α), *where $\alpha < 1$. Then the universal cover of M is homeomorphic to $\mathbb{R}^3$.*

For example, a combable group G whose combing σ is such that each word $\sigma(g)$ is geodesic has a finite presentation satisfying condition ID($\frac{1}{2}$). Every finitely presented group which possesses an almost convex Cayley graph, in the sense of Cannon [Ca], has an ID($\frac{1}{2}$) presentation [G1] (see also §4 below). Since it is known that every Nil group has at least one almost convex Cayley graph [Sp1], it follows that Nil groups have ID($\frac{1}{2}$) presentations. Another argument, proving that Nil groups have ID($\frac{3}{4}$) presentations, appears in [G1].

The difficulty with these conditions of course is that they are not invariant under change of generators.

We should also cite additional work in connection with Casson's theorem [P], [Br], [St2].

4. Relation with peak reduction algorithms

In this section, G will denote a finitely presented group with finite set of semigroup generators A and associated Cayley graph Γ.

Definition. *Let $\mu : G \to \mathbb{N}$ be a function such that $S = \{g \in G \mid \mu(g) = 0\}$ is a finite subgroup of G with $S \subset A$. Let $\mathcal{P}$ be a finite presentation for G with generators A and such that $\mathcal{P}$ contains all cyclic conjugates of its relators and their inverses and, in addition, $\mathcal{P}$ contains the group table for the finite group S. We say that $\mathcal{P}$ admits a peak reduction algorithm with respect to the function μ if the following condition holds: if $w \in A^*$ is such that $\mu(\bar{w}) \leq \mu(\bar{w}a) > \mu(\bar{w}aa')$ for some pair of generators $a, a' \in A$, then there is a relator of $\mathcal{P}$ of the form $aa' = a_1 a_2 \ldots a_k$, with $a_i \in A$, such that $\mu(\bar{w}a_1 \ldots a_i) < \mu(\bar{w}a)$ for all $1 \leq i \leq k$.*

Define a function $f_\mu : \mathbb{N} \to \mathbb{N}$ by $f_\mu(n) = \sup_{|g| \leq n} \mu(g)$.

Theorem 4.1. [G1] *Suppose that $\mathcal{P}$ admits a peak reduction algorithm for the function $\mu : G \to \mathbb{N}$. If M denotes the length of the longest relator of $\mathcal{P}$, then*

4.1.1. the function $n \mapsto M f_\mu(n) + \frac{n}{2}$ is an isodiametric function for $\mathcal{P}$, and

4.1.2. the function $n \mapsto n \cdot M^{f_\mu(n)+1}$ is an isoperimetric function for $\mathcal{P}$.

We shall now give some examples of peak reduction algorithms. With G, A, Γ as above, let $\mathcal{P}_N$ be the finite presentation with generators A and relators all words $w \in A^*$ with $\bar{w} = 1$ and $\ell(w) \leq N$. Note that it follows from the fact that G is finitely presented that $\mathcal{P}_N$ is a presentation of G for all N sufficiently large. We set B_n and S_n to be the set of vertices in the ball and sphere of radius n at the identity element in Γ.

We recall [Ca] that Γ is called almost convex iff for all n and for all pairs of points $x, y \in S_n$ which are joined by a path of length at most 3 in Γ there is a path in B_n joining these points of bounded length (where the bound is independent of $n, x,$ and y).

Proposition 4.2. [G1] *The Cayley graph Γ is almost convex if and only if there exists $N > 0$ such that $\mathcal{P}_N$ satisfies peak reduction for the function $\mu(g) = |g|$.*

The proof is not difficult from the definitions.

Corollary 4.3. *If G has an almost convex Cayley graph, then it has a linear isodiametric function and an exponential isoperimetric function.*   □

Theorem 4.4. *The groups $\mathrm{Aut}(F)$ and $\mathrm{Out}(F)$, where F is a finitely generated free group, have exponential isodiametric functions and isoperimetric functions of the form $n \mapsto A^{B^n}$.*

This is a consequence of results of Whitehead, Higgins and Lyndon, and McCool on the automorphism group of a finitely generated free group. For instance, for the group $\mathrm{Aut}(F)$ one chooses a free basis $x_1, x_2, \ldots, x_r$ for F and one takes $\mu(\phi) = \sum_{i=1}^{r} L(\phi(x_i)) - r$. Here $L(w)$, for a word w in F, is the length of a cyclically reduced word conjugate to w. The generators for $\mathrm{Aut}(F)$ are taken to be the Whitehead automorphisms [LS]. The function f_μ is seen to grow exponentially with n. McCool's algorithm [Mc] is a peak reduction algorithm for these data, so the assertion for $\mathrm{Aut}(F)$ follows from Theorem 4.1. The argument for $\mathrm{Out}(F)$ is similar.

Remark. These results for $\mathrm{Aut}(F)$ and $\mathrm{Out}(F)$ are surely not best possible. It is an open question whether $\mathrm{Out}(F)$ is automatic; if this were true, then the quadratic isoperimetric inequality would hold. In this connection, we have shown (unpublished) that neither $\mathrm{Aut}(F)$ for $\mathrm{rank}(F) \geq 3$ nor $\mathrm{Out}(F)$ for $\mathrm{rank}(F) \geq 4$ can act properly discontinuously and cocompactly on a geodesic metric space which satisfies Gromov's condition CAT(0). The situation for $\mathrm{Out}(F)$ when $\mathrm{rank}(F) = 3$ is still open.

Theorem 4.5. *$\mathrm{SL}_3(\mathbb{Z})$ has an exponential isodiametric function and an isoperimetric function of the form $n \mapsto A^{B^n}$.*

This follows from a result of Nielsen's [N], that the group $\mathrm{SL}_3(\mathbb{Z})$ satisfies a peak reduction algorithm for the function μ given by $\mu(x) = (\sum x_{ij}^2) - 3$, for $x \in \mathrm{SL}_3(\mathbb{Z})$. The generators here are the elementary transvections $E_{ij}(1)$ and the signed permutation matrices. In this case the function f_μ grows exponentially.

Remark. It follows from results of [ECHLPT] that any isoperimetric function for $\mathrm{SL}_3(\mathbb{Z})$ must grow at least exponentially. Thus from Theorem 4.5 we deduce that the Dehn function for a finite presentation of $\mathrm{SL}_3(\mathbb{Z})$ has somewhere between exponential and twice-iterated exponential growth. Which, if either, is it?

Question. Can Nielsen's argument for $\mathrm{SL}_3(\mathbb{Z})$ be generalized to a peak reduction algorithm for $\mathrm{SL}_n(\mathbb{Z})$? The answer is surely 'yes', but it seems this has never been written down (compare [Mi, §10] where a related result is established).[1] Does $\mathrm{SL}_3(\mathbb{Z})$ have a linear isodiametric function?

---

[1] We have in the meantime received the preprint [Ka] which contains the peak reduction lemma for the general linear groups.

5. Lower bounds for isoperimetric functions

The methods of this section for establishing lower bounds for the Dehn function of a finite presentation are due to [BMS] (other methods for finding lower bounds can be found in [G2]). We shall prove that the Dehn function of the free nilpotent group on $p \geq 2$ generators of class c grows at least as fast as a polynomial of degree $c + 1$. Since it is known that every finitely generated nilpotent group has a polynomial isoperimetric function [G1], it follows that arbitrary high degree polynomial growth is exhibited by these free nilpotent groups as $c \to \infty$.

Let $\mathcal{P} = \langle x_1, x_2, \ldots, x_p \mid R_1, R_2, \ldots, R_q \rangle$ and let F be the free group freely generated by the generators $x_1, x_2, \ldots, x_p$ and let $N \triangleleft F$ be the normal closure of the relators. We let $G = G(\mathcal{P}) = F/N$ as earlier.

Proposition 5.1. *The group* $N/[F, N]$ *is a finitely generated abelian group.*

Proof. The identity $R_i^u = [u, R_i]R_i \in [F, N]R_i$ shows that the cosets of the relators R_i generate the factor group $N/[F, N]$. Since $[F, N] \supset [N, N]$, this factor group is abelian, and consequently it is a finitely generated abelian group.

Definition. *Let* $V = \mathbb{Q} \otimes N/[F, N]$, *considered as a finitely generated vector space over* $\mathbb{Q}$. *If* $v_1, v_2, \ldots v_d$ *is a basis for* V, *we define the* ℓ_1-*norm* $|v|_1$ *of a vector* $v \in V$ *with respect to this basis to be* $\sum_{i=1}^{d} |a_i|$, *where* $v = \sum_{i=1}^{d} a_i v_i$, $a_i \in \mathbb{Q}$.
If $w \in N$, *then we define* $|w|_1$ *to be the* ℓ_1-*norm of* $1 \otimes [w]$, *where* $[w]$ *is the coset* $w[F, N] \in N/[F, N]$.

Theorem 5.2. *With the notations above, there is a constant* $C \geq 0$ *so that for all* $w \in N$ *we have*
$$|w|_1 \leq C \operatorname{Area}_{\mathcal{P}}(w).$$

Proof. Let $C = \max_{1 \leq i \leq q} |R_i|_1$. This is the number C of the theorem. Suppose now that $w \in N$, so $w = \prod_{j=1}^{k} R_{i_j}^{\epsilon_j u_j}$, where $\epsilon_j = \pm 1$ and $u_j \in F$ and where $k = \operatorname{Area}_{\mathcal{P}}(w)$. Observe that since $R_{i_j}^{u_j} \in [F, N]R_{i_j}$, we have $[R_{i_j}^{\epsilon_j u_j}] = \epsilon_j[R_{i_j}]$ in V. From this it follows that $|w|_1 = |\sum_{j=1}^{k} \epsilon_j[R_{i_j}]|_1 \leq Ck \leq C \operatorname{Area}_{\mathcal{P}}(w)$. This completes the proof.

Remark. If we change the basis of V above and calculate the ℓ_1-norm with respect to the new basis, the effect is to change the constant C in Theorem 5.2.

Next we recall some facts about nilpotent groups. A central series for a group G is sequence of subgroups
$$H_n < H_{n-1} < \ldots < H_0 = G$$
so that $[G, H_i] < H_{i+1}$ for all i. The group G is called nilpotent if it has such a central series with $H_n = 1$ for some n, and the minimum such number n for

all central series is called the class of nilpotence. For example, a nontrivial abelian group has class 1 and the Heisenberg group has class 2. The lower central series $\{G_n, n \geq 0\}$ for any group G is defined inductively by $G_0 = G$, $G_{n+1} = [G, G_n]$. One has that $G_i < H_i$ for any central series H_i as above, so the lower central series descends at least as fast as any central series for G.

In particular we can apply these notions to the free group F freely generated by $x_1, x_2, \ldots, x_p$, where $p \geq 2$, to get the lower central series $\{F_n\}$ of the free group. The group F/F_c is called the free nilpotent group on p generators of class c. It is a standard result that the normal subgroup $F_c \lhd F$ is generated by all left normed commutators of length $c+1$, $\mathrm{ad}(u_1) \circ \mathrm{ad}(u_2) \circ \ldots \circ \mathrm{ad}(u_c)(u_{c+1})$, where $u_i \in F$. Here $\mathrm{ad}(u)(v) = [u, v]$. As an example, using this observation it is easy to see that the free nilpotent group on 2 generators of class 2 is the Heisenberg group.

Theorem 5.3. *The free nilpotent group on $p \geq 2$ generators of class $c \geq 1$ has the property that its Dehn function grows at least as fast as a polynomial of degree $c + 1$.*

Proof. [2]   Since the free nilpotent group on $p \geq 2$ generators and class c retracts to that on 2 generators and class c, it suffices to prove the result for 2 generators. Let $F = F(a, b)$ be the free group freely generated by a, b and let $w_n = \mathrm{ad}(a^n)^{(c)}(b^n) \in F_c$. One checks that $\ell(w_n)$ grows linearly with n. However, when $[w_n]$ is considered in $V = \mathbb{Q} \otimes F_c/[F, F_c] = \mathbb{Q} \otimes F_c/F_{c+1}$, one has by multilinearity $[w_n] = n^{c+1}[\mathrm{ad}(a)^{(c)}(b)]$. But it is known that the Engel element $\mathrm{ad}(a)^{(c)}(b)$ of the free abelian group F_c/F_{c+1} is an element of a $\mathbb{Z}$-basis [MKS, §5.7 Problem 4], so $[\mathrm{ad}(a)^{(c)}(b)] \neq 0$ in V. It follows that $|w_n|_1 = n^{c+1}|[\mathrm{ad}(a)^{(c)}(b)]|_1 \neq 0$, so $|w_n|_1$ grows like a polynomial in n of degree $c + 1$. It follows from Theorem 5.2 that $\mathrm{Area}_{\mathcal{P}}(w_n)$ grows at least as fast as a polynomial in n of degree $c + 1$, where $\mathcal{P}$ is a finite presentation for F/F_c. Since $\ell(w_n)$ is linear in n, it follows that the Dehn function for F/F_c must grow at least as fast as a polynomial of degree $c + 1$. This completes the proof.

Remark. If $N < [F, F]$ above, then $N/[F, N] \cong H_2(G, \mathbb{Z})$, as one sees from Hopf's formula. In this case the vector space V is $H_2(G, \mathbb{Q})$.

Remark. Taking $p = 2$ and $c = 2$ in Theorem 5.3, we recover the result of [ECHLPT] and [G2] that the Dehn function for the 3-dimensional integral Heisenberg group grows at least as fast as a cubic polynomial. The next result shows that this result is optimal (other proofs that the 3-dimensional integral Heisenberg group has a cubic polynomial for its Dehn function are given in [ECHLPT] and [G2]).

---

[2] H. Short told me the statement of Theorem 5.3 at the Sussex conference, from which I worked out the proof given here. Baumslag, Miller, and Short wrote me subsequently that this argument was one of several they had in mind.

Proposition 5.4. *The Dehn function for the 3-dimensional integral Heisenberg group H grows like a cubic polynomial.*

Proof. We have already shown that the Dehn function grows at least as fast as a cubic polynomial. We shall obtain now a cubic polynomial upper bound. A presentation for H is $\mathcal{P} = \langle x, y, t \mid x^t = xy, y^t = y, xy = yx \rangle$. Let $\mathcal{Q} = \langle x, y, t \mid x^t = xy, y^t = y \rangle$ and let $\mathcal{R} = \langle x, y \mid xy = yx \rangle$. Observe that $\mathcal{Q}$ is a presentation for the split extension $F(x, y) \rtimes_\phi \mathbb{Z}$ of the free group $F(x, y)$, where $\phi(x) = xy, \phi(y) = y$.

Let w be a word in the generators of $\mathcal{P}$ with with $\ell(w) = n$ and such that $\bar{w} = 1$ in H. We shall find a van Kampen diagram for w in two steps. First, using only the relations of $\mathcal{Q}$, we find a sequence of cyclic words $w = w_0, w_1, \ldots, w_{n-1} = w'$, where each is obtained from the preceding by at most a single t-reduction (viewing t as the stable letter in the HNN extension $F(x, y) \rtimes_\phi \mathbb{Z}$ with base group $F(x, y)$), until one runs out of t-letters. If ℓ_x, ℓ_y, ℓ_t denote respectively the number of letters $x^\pm, y^\pm, t^\pm$ in a free word, we see inductively that $\ell_x(w_i) \le \ell_x(w)$, $\ell_y(w_i) \le \ell_x(w_{i-1}) + \ell_y(w_{i-1})$, and $\ell_t(w_i) \le \max(\ell_t(w_{i-1}) - 2, 0)$. It follows that $\ell_y(w_i) \le i\ell_x(w) + \ell_y(w) \le (i+1)n$, and there is an annular diagram A_i in $\mathcal{Q}$ connecting w_{i-1} with w_i of area $\text{Area}(A_i) \le \ell(w_{i-1}) \le in$. If we fit these annular diagrams together, we obtain an annular diagram D_1 in $\mathcal{Q}$ with boundary components labelled w and w' such that $\text{Area}(D_1) \le \sum_{i=1}^{n-1} in = O(n^3)$

Since $\ell_x(w') \le n$, $\ell_y(w') \le n^2$, and $\ell_t(w') = 0$, we can find a disc diagram D_2 for w' in $\mathcal{R}$ with $\text{Area}(D_2) \le n^3$. If we fit D_1 and D_2 together along their common boundary component labelled w', we obtain a disc diagram D for w with $\text{Area}(D) \le O(n^3) + n^3 = O(n^3)$. This completes the proof of the proposition.

The same method as in Theorem 5.3 suffices to prove the following result.

Proposition 5.5. *If G is a finitely generated nilpotent group of class c given by the exact sequence $1 \to N \to F \to G \to 1$, with F finitely generated and free (so $F_c < N$), and if the canonical map $F_c \to N/[F, N]$ has infinite image, then the Dehn function for G grows at least as fast as a polynomial of degree $c + 1$.* □

Example. The $(2n+1)$-dimensional integral Heisenberg group $\mathfrak{H}_{2n+1}$ is given by the presentation

$$\langle x_1, x_2, \ldots, x_n, y_1, y_2, \ldots, y_n \mid [x_i, x_j] = 1, [y_i, y_j] = 1 \text{ for all } i, j,$$
$$[x_k, y_l] = 1 \text{ for all } k \ne l, \quad [x_m, y_m] = z \text{ for all } m, \quad z \text{ central }\rangle.$$

It is more convenient to use another presentation for $\mathfrak{H}_{2n+1}$ obtained by Tietze transformations from the preceding by eliminating the central generator z. This new presentation has generators $x_1, x_2, \ldots, x_n, y_1, y_2, \ldots y_n$ freely generating the free group F of rank $2n$. The normal subgroup of relations N is contained in $[F, F] = F_1$ for this second presentation, and since $\mathfrak{H}_{2n+1}$ is nilpotent of class 2, we have $F_2 \subset N$. We have then the next result.

Proposition 5.6. *With the notations preceding we have*

5.6.1. if $n \geq 2$, the canonical homomorphism

$$F_2/F_3 \to N/[F, N] = H_2(\mathfrak{H}_{2n+1}, \mathbb{Z})$$

is the zero map, whereas

5.6.2. if $n = 1$, then we have $F_2/F_3 \xrightarrow{\cong} H_2(\mathfrak{H}_3, \mathbb{Z})$.

Proof. The second assertion (5.6.2) follows from earlier remarks since $\mathfrak{H}_3$ is the free nilpotent group of class 2 on 2 generators. We proceed then to the proof of (5.6.1).

The group F_2/F_3 is generated by elements $[u, [v, w]]$ where each of u, v, w is in the set $\{x_i, y_i;\ 1 \leq i \leq n\}$. Since we have $[x_i, y_j] \in N$ for $i \neq j$ and since $[x_k, x_l]$, $[y_k, y_l] \in N$ for all k, l, it follows that we have $[u, [x_i, y_j]] \in [F, N]$ for $i \neq j$ and $[u, [x_k, x_l]] \in [F, N]$, $[u, [y_k, y_l]] \in [F, N]$ for all k, l, where $u \in \{x_i, y_i;\ 1 \leq i \leq n\}$.

It remains to prove that $[x_j, [x_i, y_i]] \in [F, N]$ and $[y_j, [x_i, y_i]] \in [F, N]$ for all i, j. We shall prove the first assertion, since the second follows symmetrically. If $j \neq i$, this assertion is a consequence of an identity attributed variously to E. Witt or P. Hall,

$$[b, [a^{-1}, c]]^a [a, [c^{-1}, b]]^c [c, [b^{-1}, a]]^b = 1,$$

for all elements a, b, c in a group. We remind the reader here that our convention for commutators is $[a, b] = aba^{-1}b^{-1}$ and $a^b = bab^{-1}$. If we substitute $b = x_j, a = x_i^{-1}, c = y_i$ in this identity, we find that two terms of the product are in $[F, N]$, whence the third, a conjugate of $[x_j, [x_i, y_i]]$, is also in $[F, N]$. It remains then to prove that $[x_i, [x_i, y_i]] \in [F, N]$. Since $n \geq 2$, there is an index $j \neq i$. Observe first that $[x_i x_j, y_i y_j^{-1}] \in N$, as one sees by an elementary computation in $\mathfrak{H}_{2n+1}$, so we have $[x_i, [x_i x_j, y_i y_j^{-1}]] \in [F, N]$. The commutator $[x_i x_j, y_i y_j^{-1}]$ can be expanded as the product of four commutators, making use of the relation $[a, bc] = [a, b][a, c]^b$; we obtain

$$[x_i x_j, y_i y_j^{-1}] = [x_j, y_i]^{x_i}[x_i, y_i][x_j, y_j^{-1}]^{y_i x_i}[x_i, y_j^{-1}]^{y_i}.$$

When we expand the expression $[x_i, [x_i x_j, y_i y_j^{-1}]]$ and make use of the commutator identities (and the fact that $F_2 < N$, so $F_3 < [F, N]$), we obtain that this class in $N/[F, N]$ is that of the product of four classes, those of $[x_i, [x_j, y_i]]$, $[x_i, [x_i, y_i]]$, $[x_i, [x_j, y_j^{-1}]]$, and $[x_i, [x_i, y_j^{-1}]]$. The first, third, and fourth terms have already been shown to be in $[F, N]$. Since the product lies in $[F, N]$, it follows that the second term, $[x_i, [x_i, y_i]]$, is also in $[F, N]$. This completes the proof of the Proposition.

Remark. The proposition just proved shows that Thurston's assertion, that $\mathfrak{H}_{2n+1}$ satisfies the quadratic isoperimetric inequality for $n \geq 2$, is consistent with Proposition 5.5: if the homomorphism $F_2/F_3 \to N/[F, N]$ in (5.6.1) had had infinite image, then the Dehn function for this group would have been a cubic polynomial and not quadratic.

Remark. It is proved in [G1] that a finitely generated nilpotent group has a polynomial isoperimetric function of degree 2^h, where h is the Hirsch number. The bound on the degree was improved in [Co] to $2 \cdot 3^c$, where c is the class of nilpotence. We do not know an example of a finitely generated nilpotent group where there does not exist an isoperimetric polynomial of degree $c + 1$.

Acknowledgments. I wish to thank Alexander Ol'shanskii, Hamish Short, and John Stallings for their helpful criticisms.

References

[ABC]   J. Alonso, T. Brady, D. Cooper, V. Ferlini, M. Lustig, M. Mihalik, M. Shapiro, and H. Short, *Notes on word hyperbolic groups,* in: Group Theory from a Geometrical Viewpoint (E. Ghys, A. Haefliger, and A. Verjovsky, eds.), World Scientific, 1991, pp. 3–63.

[Al]   J. Alonso, *Inégalités isopérimétriques et quasi-isométries,* C. R. Acad. Sci. Paris, Série 1 **311** (1991), pp. 761–764.

[Bd]   M. Bridson, *Combing the fundamental group of a Haken 3-manifold,* preprint, Princeton University, 1992.

[BF]   M. Bestvina and M. Feighn, *A combination theorem for negatively curved groups,* Jour. Diff. Geom. **35** (1992), pp. 85–101.

[BGSS]   G. Baumslag, S. M. Gersten, M. Shapiro, and H. Short, *Automatic groups and amalgams,* Jour. Pure Appl. Alg. **76** (1991), pp. 229–316.

[BMS]   G. Baumslag, C. F. Miller III, and H. Short, *Isoperimetric inequalities and the homology of groups,* preprint, CUNY, 1992.

[Br]   S. Brick, *Filtrations of universal covers and a property of groups,* preprint, University of California, Berkeley, 1991.

[Ca]   J. W. Cannon, *Almost convex groups,* Geometriae Dedicata **22** (1987), pp. 197–210.

[C]   D. E. Cohen, *Isodiametric and isoperimetric inequalities for group presentations,* Internat. J. Algebra and Computing **1** (1991), pp. 315–320.

[Co]   G. Conner, Ph. D. Thesis, Univ. of Utah, 1992.

[ECHLPT] J. W. Cannon, D. B. A. Epstein, D. Holt, S. Levy, M. Paterson, and W. P. Thurston, *Word processing in groups,* Bartlett and Jones, Boston, 1992.

[FHL]   W. J. Floyd, A. H. M. Hoare, and R. C. Lyndon, *The word problem for geometrically finite groups,* Geometriae Dedicata **20** (1986), pp. 201–207.

[G1]   S. M. Gersten, *Isodiametric and isoperimetric inequalities in group extensions*, preprint, Univ. of Utah, 1991.

[G2]   S. M. Gersten, *Dehn functions and ℓ_1-norms of finite presentations*, in: Proceedings of a Workshop on Algorithmic Problems (G. Baumslag and C. F. Miller III, eds.), MSRI series 23, Springer, 1991.

[G3]   S. M. Gersten, *Bounded cohomology and combings of groups*, preprint, Univ. of Utah, 1991.

[G4]   S. M. Gersten, *The double exponential theorem for isodiametric and isoperimetric functions*, Internat. J. Algebra and Comput. 1 (1991), pp. 321–327.

[Gh]   E. Ghys, *Les groupes hyperboliques*, in: Sem. Bourbaki (1989–1990), no. 722.

[GH]   E. Ghys and P. de la Harpe, *Sur les groupes hyperboliques d'après Mikhael Gromov*, Birkhäuser, 1990.

[GLP]   M. Gromov, J. Lafontaine and P. Pansu, *Structures métriques pour les variétés riemanniennes*, Cedic/F. Nathan, Paris, 1981.

[Gr]   M. Gromov, *Hyperbolic groups*, in: Essays in group theory (S. M. Gersten, ed.), MSRI series 8, Springer, 1987.

[Gr2]   M. Gromov, *Asymptotic invariants of infinite groups*, preprint, IHES, 1992.

[Ka]   S. Kalajdžievski, *PRL for general linear groups*, preprint, Univ. of Manitoba, 1992.

[LS]   R. C. Lyndon and P. E. Schupp, *Combinatorial group theory*, Springer, 1977.

[Mc]   J. McCool, *A presentation for the automorphism group of a free group of finite rank*, Jour. London Math. Soc. 8 (1974), pp. 259–266.

[Mi]   J. Milnor, *Algebraic K-Theory*, Annals of Math. Study 72, Princeton Univ. Press, 1971.

[MKS]   W. Magnus, A. Karrass, and D. Solitar, *Combinatorial group theory*, J. Wiley, 1966.

[N]   J. Nielsen, *Die Gruppe der Dreidimensionalen Gittertransformationen*, in: Collected Mathematical Papers of Jakob Nielsen, Vol. 1 (1913–1932), Birkhäuser, 1986, pp. 147–173.

[Ol]   A. Yu. Ol'shanskii, *Hyperbolicity of groups with subquadratic isoperimetric inequality*, Internat. J. Algebra and Comput. 1 (1991), pp. 281–289.

[P]   V. Poénaru, *Geometry "á la Gromov" for the fundamental group of a closed 3-manifold M^3 and the simple connectivity at infinity of $\widetilde{M}^3$*, preprint, Univ. de Paris-Sud, Orsay, 1990.

[Sh]   H. Short, *Groups and combings*, preprint, ENS Lyon, July 1990.

[Sp1]   M. Shapiro, *A geometric approach to the almost convexity and growth of some nilpotent groups*, Math. Ann. (1989), pp. 601–624.

[Sp2]   M. Shapiro, *Deterministic and non-deterministic asynchronous automatic structures*, preprint, Ohio State Univ., June 14, 1991.

[St1]   J. Stallings, *Topology of finite graphs*, Inv. Math. 71 (1983), pp. 551–565.

[St2]   J. Stallings, *Brick's quasi simple filtrations for groups and 3-manifolds*, these Proceedings.

[SG]   J. Stallings and S. M. Gersten, *Casson's idea about 3-manifolds whose universal cover is $\mathbb{R}^3$*, to appear in Internat. J. Algebra and Computing.

[Th]   W. P. Thurston, *Three-dimensional manifolds, Kleinian groups and hyperbolic geometry*, Bull. Amer. Math. Soc. 6 (1982), pp. 357–381.

On Hilbert's Metric for Simplices

Pierre de la Harpe

Section de Mathématiques, Université de Genève, 2-4 rue du Lièvre, C.P. 240,
1211 Genève 24, Switzerland.

Abstract. For any bounded convex open subset C of a finite dimensional real vector space, we review the canonical Hilbert metric defined on C and we investigate the corresponding group of isometries. In case C is an open 2-simplex S, we show that the resulting space is isometric to $\mathbb{R}^2$ with a norm such that the unit ball is a regular hexagon, and that the central symmetry in this plane corresponds to the quadratic transformation associated to S. Finally, we discuss briefly Hilbert's metric for symmetric spaces and we state some open problems.

1. Generalities on Hilbert metrics

The first proposition below comes from a letter of D. Hilbert to F. Klein [Hil]. It is discussed in several other places, such as sections 28, 29 and 50 of [BuK], and chapter 18 of [Bu1], and [Bea]. There are also nice applications of Hilbert metrics to the classical Perron-Frobenius Theorem [Sae], [KoP] and to various generalizations in functional analysis [Bir], [Bus].

Let V be a real affine space, assumed here to be finite dimensional (except in Remark 3.3), and let C be a *non empty bounded convex open* subset of V. We want to define a metric on C which, in the special case where C is the open unit disc of the complex plane, gives the projective model of the hyperbolic plane (sometimes called the "Klein model").

Let $x, y \in C$. If $x = y$, one sets obviously $d(x, y) = 0$. Otherwise, the well defined affine line $\ell_{x,y} \subset V$ containing x and y cuts the boundary of C in two points, say u on the side of x and v on the side of y; see Figure 1. One sets

$$d(x, y) = \log [x, y, v, u] = \log \left(\frac{v - x}{v - y} : \frac{u - x}{u - y} \right),$$

where $\dfrac{v-x}{v-y}$ denotes the ratio $\dfrac{f(v)-f(x)}{f(v)-f(y)}$ for some (hence all) affine
bijection f from $\ell_{x,y}$ onto $\mathbb{R}$. (In particular this definition makes sense *without*
any metric on V; indeed $[x, y, v, u]$ is invariant by any *projective* isomorphism
of the projective line $\ell_{x,y} \cup \{\infty\}$.) Notations are such that $[x, y, v, u] > 1$, so
that $d(x, y) > 0$.

Proposition 1. (Hilbert) *Notations being as above, d is a metric on the*
convex set C.

Proof. We check the triangle inequality for three points x, y, z in C. If these
are on a line, say with z between x and y, it is straightforward to check the
equality $d(x, y) = d(x, z) + d(z, y)$. If x, y, z are not collinear, we introduce
in the plane which contains them the following points, as in Figure 1 (where
the curve γ indicates the intersection of the boundary of C with the plane
containing x, y and z):

> u and v, where $\ell_{x,y}$ cuts γ,
> a and c, where $\ell_{z,y}$ cuts γ,
> b and d, where $\ell_{x,z}$ cuts γ,
> p, the intersection of $\ell_{a,b}$ and $\ell_{c,d}$ (this p may be at infinity),
> $u' = \ell_{a,b} \cap \ell_{x,y}$, $z' = \ell_{p,z} \cap \ell_{x,y}$, $v' = \ell_{d,c} \cap \ell_{x,y}$.

One has on one hand
$$[x, z, d, b] = [x, z', v', u'],$$
$$[z, y, c, a] = [z', y, v', u'],$$

by projective invariance of the cross-ratio (see for example Lemma 6.5.3 in
[Ber]). On the other hand

$$[x, y, v', u'] = [x, z', v', u'][z', y, v', u'],$$
$$[x, y, v, u] \leq [x, y, v', u'],$$

by straightforward computations. It follows that

$$d(x, y) \leq \log{[x, z', v', u']} + \log{[z', y, v', u']} = d(x, z) + d(z, y),$$

as claimed.   □

In case V is of dimension 1, it is easy to check that the metric d makes C
isometric to the Euclidean line. From now on, we *assume that the dimension*
of V is at least 2.
Let C and V be as above. The closure $\overline{C}$ of C is compact and convex. Two
distinct points $p, q \in \overline{C}$ define a closed segment, denoted by $[p, q]$, and an open
segment, denoted by $]p, q[$, both in $\overline{C}$. A *face* of $\overline{C}$ is a convex subset F of $\overline{C}$
such that, for any pair (p, q) of distinct points of $\overline{C}$ such that $]p, q[\cap F \neq \varnothing$,
one has $[p, q] \subset F$. A face is *proper* if it is neither empty nor $\overline{C}$ itself. The
relative interior of a face F of $\overline{C}$ is the interior of F in the smallest affine

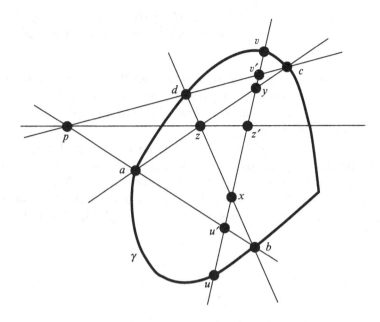

Figure 1: The triangular inequality.

subspace of V containing F. It is known that faces of $\overline{C}$ are closed, that proper faces of $\overline{C}$ are in the boundary $\partial\overline{C}$ of $\overline{C}$, and that the relative interiors of the proper faces of $\overline{C}$ constitue a partition of $\partial\overline{C}$ (see e.g. [Brø], Theorems 5.1, 5.3 and 5.7).

Proposition 2. *Let $C \subset V$ be as above (with $\dim(V) \geq 2$). Let x, y be two distinct points of C, let u, v be the points where $\ell_{x,y}$ meets $\partial\overline{C}$, and let F, G be the faces of $\overline{C}$ whose relative interiors contain u, v, respectively. The following are equivalent.*

  (i) There exists $z \in C$ such that x, y, z are not collinear and such that $d(x,y) = d(x,z) + d(z,y)$.

  (ii) There exist open segments I, J such that $u \in I \subset F$, $v \in J \subset G$ and such that I, J span a 2-dimensional affine plane of V.

Proof. This is a straightfoward consequence of the proof of Proposition 1. Indeed, with the same notations as in this proof, (i) above holds if and only if there exists $z \in C - \ell_{x,y}$ such that $u' = u$ and $v' = v$, and this in turn is equivalent to (ii). □

Recall that $\overline{C}$ is said to be *strictly convex* if all its proper faces are reduced to points. Proposition 2 implies obviously the following.

Corollary. *Assume that $\overline{C}$ is strictly convex, or more generally that all but possibly one of its proper faces are reduced to points. Then C is a geodesic* *space for the Hilbert metric; more precisely, given any pair (p, q) of distinct points in C, there is a unique geodesic segment between p and q, and this is the straight segment $[p, q]$.*

By similar arguments, one shows: (i) that balls for Hilbert metrics are always convex and (ii) that these balls are strictly convex if C fulfils the hypothesis of the previous Corollary. The same holds for horoballs, these being easily defined in case $\partial \overline{C}$ is smooth.

Here are two examples to which the Corollary does *not* apply. Let first S_2 be an open triangle in a plane and let x, y be two distinct points in S_2. Denote by $I(x, y)$ the set of those points $z \in S_2$ such that $d(x, y) = d(x, z) + d(z, y)$. If the line $\ell_{x,y}$ through x and y contains a vertex of $\overline{S_2}$, then $I(x, y) = [x, y]$ by Proposition 2. Otherwise, one has $[x, y] \subsetneq I(x, y)$, and $I(x, y)$ is as in Figure 2.

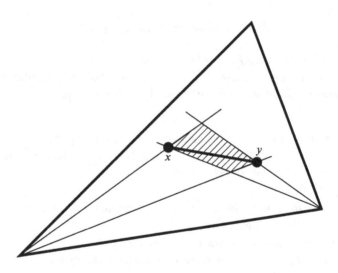

Figure 2: The set $I(x, y)$.

Now let S_3 be an open tetrahedron in $\mathbb{R}^3$. Let u, v be two points on the interiors of two opposite edges of $\overline{S_3}$, and let x, y be two distinct points in $]u, v[$. With the notations of Proposition 2, the faces F and G are the closed edges of $\overline{S_3}$ containing u and v. As these edges are skew, Proposition 2 shows that there exists a unique geodesic segment joining x and y. For generic points $x', y' \in S_3$, we leave it to the reader to define a set $I(x', y')$ as in the previous example and to check that $I(x', y')$ is a non degenerate polytope.

Let again $C \subset V$ be as in the beginning of the present section; we assume moreover that V is an open cell inside some projective space $\overline{V}$ (see e.g. §I.C in [Sam]). Let $Isom(C)$ denote the group of all *isometries* of C for the Hilbert metric. Let $Coll(\overline{V})$ denote the group of all *collineations* (or projective linear isomorphisms) of $\overline{V}$ and let $Coll(C)$ denote the subgroup of these $g \in Coll(\overline{V})$ for which $g(C) = C$. Observe that $Coll(C)$ is a closed subgroup of $Coll(\overline{V}) \approx PGL_{n+1}(\mathbb{R})$, so that $Coll(C)$ is naturally a Lie group. One has

$$Coll(C) \subset Isom(C),$$

by projective invariance of the cross-ratio. The following Proposition appears in (29.1) of [BuK] (stated in the particular case where $\dim(V) = 2$). By a *line* in C, we mean a subset of the form $]u, v[= \ell_{u,v} \cap C$ where the points $u, v \in \partial\overline{C}$ are such that $]u, v[\not\subset \partial\overline{C}$.

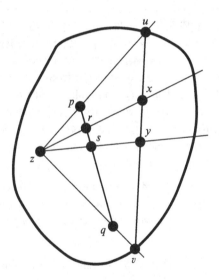

Figure 3: Equal cross-ratios.

Lemma. *Suppose $\overline{C}$ is strictly convex, and consider $g \in Isom(C)$. Then:*

(i) The image by g of a line in C is again a line in C.

(ii) The map g extends to a homeomorphism of $\overline{C}$.

(iii) Cross-ratios of collinear points inside C are preserved by g.

Proof. Claim (i) follows from the previous corollary.

For (ii), define the extension $g(u)$ of g to a point $u \in \partial\overline{C}$ as follows. Choose $x \in C$, let $f : [0, \infty[\to [x, u[$ denote the isometric parametrization of the ray (i.e. half-line) from x towards u, so that gf is a parametrization of the ray $g([x, u[)$, and set $g(u) = \lim_{t\to\infty} gf(t)$. Let us check that this definition of $g(u)$ does not depend on x.

For any other choice $x' \in C$ and for the corresponding $f' : [0, \infty[\to [x', u[$, set $g'(u) = \lim_{t\to\infty} gf'(t)$. It is easy to check that the Hausdorff distance between $[x, u[$ and $[x', u[$ is finite. If one had $g(u) \neq g'(u)$, the Hausdorff distance between $[g(x), g(u)[= g([x, u[)$ and $[g(x'), g'(u)[= g([x', u[)$ would be infinite (note that strict convexity is crucial here, as illustrated by Figure 4), and this would be absurd. Hence $g'(u) = g(u)$.

Let $(u_j)_{j\geq 1}$ be a sequence of points in $\overline{C}$ converging towards some $u \in \partial\overline{C}$. Denote by f_j the isometric parametrization $[0, d(x, u_j)] \to [x, u_j]$, if $u_j \in C$, or $[0, \infty[\to [x, u_j[$, if $u_j \in \partial\overline{C}$, and denote by f the isometric parametrization $[0, \infty[\to [x, u[$. One has clearly $\lim_{j\to\infty} d(f_j(1), f(1)) = 0$, thus also $\lim_{j\to\infty} d(gf_j(1), gf(1)) = 0$, and it follows that $\lim_{j\to\infty} g(u_j) = g(u)$. Thus $g : \overline{C} \to \overline{C}$ is continuous. We leave it to the reader to check that this latter map is a homeomorphism.

For (iii), consider four pairwise distinct collinear points $p, r, s, q \in C$, say with $r, s \in]p, q[$ as in Figure 3. Choose $z \in C$ such that $z \notin \ell_{p,q}$. (Remember the standing hypothesis $\dim(V) \geq 2$; the reader interested by the case $\dim(V) = 1$ will find his own argument.) Let u (respectively v) be the intersection with $\partial\overline{C}$ of the ray starting from z and containing p (resp. q). Set $x = \ell_{z,r} \cap]u, v[$ and $y = \ell_{z,s} \cap]u, v[$, so that $[r, s, q, p] = [x, y, v, u]$. As $g(r), \ldots, g(p)$ are projectively related via $g(z)$ to $g(x), \ldots, g(u)$, one has similarly $[g(r), g(s), g(q), g(p)] = [g(x), g(y), g(v), g(u)]$. As g is an isometry, one has also $[g(x), g(y), g(v), g(u)] = [x, y, v, u]$. The three previous equalities imply $[g(r), g(s), g(q), g(p)] = [r, s, p, q]$.   □

Figure 4 is related to the proof of the Lemma and shows that, when the parallel lines $\ell_{y,y'}$ converge to $\ell_{v,v'}$, one may have $\lim d(y, y') = \log[v, v', b, a] < \infty$ in case $\overline{C}$ is not strictly convex, whereas $\lim d(y, y') = \infty$ if $\overline{C}$ is strictly convex.

Proposition 3. *Let $C \subset V \subset \overline{V}$ be as above, and assume that $\overline{C}$ is strictly convex. Then*

$$Coll(C) = Isom(C).$$

Proof. Set $n = \dim(\overline{V})$. Let $\{x_0, \ldots, x_n, y_0\}$ be a projective frame in $\overline{V}$. (This means that there are linear coordinates $(\xi_0, \ldots, \xi_n)$ in the $(n + 1)$-vector space W above $\overline{V}$ such that $\{x_0, \ldots, x_n\}$ is the image in $\overline{V}$ of the corresponding linear basis and such that y_0 is the image in $\overline{V}$ of the vector $\tilde{y}_0$ with all coordinates equal to 1.) For each $k \in \{1, \ldots, n\}$, let y_k denote the intersection of the projective line L_k containing x_0, x_k and of the projective hyperplane containing $x_1, \ldots, x_{k-1}, x_{k+1}, \ldots, x_n, y_0$. It is easy to check that x_0, x_k, y_k are distinct, i.e. that they form a projective frame in L_k.

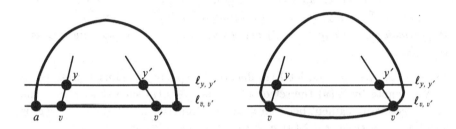

Figure 4: Rays and strict convexity.

Let us assume firstly that $y_0 \in C$. As we may choose $x_0, \ldots, x_n$ in a small neighbourhood of y_0, we may also assume that $x_0, \ldots, x_n$ are in C. It follows that $y_1, \ldots, y_n$ are also in C.

Consider now $g \in \textit{Isom}(C)$; we have to show that $g \in \textit{Coll}(C)$. As $g : C \to C$ is a bijection preserving the lines, the points $g(x_0), \ldots, g(x_n), g(y_0)$ constitute also a projective frame. Thus, there exists a unique $h \in \textit{Coll}(\overline{V})$ such that $h(x_k) = g(x_k)$ for each $k \in \{0, \ldots, n\}$ and such that $h(y_0) = g(y_0)$. (See for example Proposition 4.5.10 in [Ber].) Let $k \in \{1, \ldots, n\}$. As both g and h preserve lines, one has also $h(y_k) = g(y_k)$. As both g and h preserve cross-ratios, and as they coincide on the three distinct points x_0, x_k, y_k of the projective line L_k, they coincide on the whole of $L_k \cap C$.

Let $j, k \in \{1, \ldots, n\}$ be such that $j \neq k$, and let $L_{j,k}$ be the projective plane containing L_j and L_k. For any $x \in L_{j,k} \cap C$ such that $x \notin L_j \cup L_k$, choose two distinct lines ℓ, ℓ' which meet $L_j \cup L_k$ in four distinct points. As the images of ℓ and ℓ' by g coincide with their images by h, one has also $h(x) = g(x)$.

Similarly, by induction on the subsets of $\{1, \ldots, n\}$, one shows that h and g coincide on the whole of C. Thus one has $h \in \textit{Coll}(C)$ and the proof is complete. $\qquad\qquad\qquad\qquad\qquad\qquad\qquad\qquad\qquad\qquad\qquad\qquad\qquad$ □

More generally, I believe that $\textit{Isom}(C)$ is naturally a Lie group, and that its closed subgroup $\textit{Coll}(C)$ is also open, namely that the connected component of $\textit{Coll}(C)$ coincides with that of $\textit{Isom}(C)$.

It would be interesting to know exactly for which convex sets $C \subset V$ one has $Coll(C) = Isom(C)$. The following easy fact shows in particular that this equality holds in case C is the interior of a convex plane quadrilateral. We denote by D_n the dihedral group of order $2n$.

Proposition 4. *Let C be the interior of a convex polygon $\overline{C}$ with N vertices in a 2-dimensional plane V.*

(i) If $N \geq 4$ the groups $Coll(C)$ and $Isom(C)$ are finite.

(ii) If $N = 4$ the groups $Coll(C)$ and $Isom(C)$ are equal, and isomorphic to D_4.

(iii) If $N = 3$ the group $Coll(C)$ is isomorphic to the natural semi-direct product $\mathbb{R}^2 \rtimes D_3$, and it is a subgroup of index 2 in $Isom(C)$.

In particular the group $Isom(C)$ acts transitively on C if and only if C is a triangle.

Proof. For each $p \in C$, let $\Sigma(p)$ denote the sphere of radius 1 centered at p. We denote by $\Sigma'(p)$ the set of all points $x \in \Sigma(p)$ such that $[p, x]$ is the unique geodesic segment between p and x. Proposition 2 shows that $\Sigma'(p)$ is also the set of those $x \in \Sigma(p)$ such that $\ell_{p,x}$ meets $\partial \overline{C}$ at a vertex.

Suppose that $N \geq 4$, and let $v_1, \ldots, v_N$ be the vertices of $\overline{C}$. We denote by F the finite subset of C consisting of the intersections of two distinct lines of the form ℓ_{v_i, v_j}. If $p \in C$ is not on any of the lines ℓ_{v_i, v_j}, the set $\Sigma'(p)$ is the union of the N distinct pairs $\Sigma(p) \cap \ell_{p, v_k}$, so that $|\Sigma'(p)| = 2N$. If $p \in C$ is on exactly one of the lines ℓ_{v_i, v_j}, then $|\Sigma'(p)| = 2(N-1)$. If $p \in F$, then $|\Sigma'(p)| \leq 2(N-2)$. Thus F is invariant by the group $Isom(C)$. Moreover, $Isom(C)$ acts faithfully on the finite set $\bigcup_{p \in F} \Sigma'(p)$, so that $Isom(C)$ is finite. If $N = 4$, the set F is reduced to one point, say o, and $|\Sigma'(o)| = 4$. As C is projectively isomorphic to a square, it is obvious that $Isom(C)$ is equal to $Coll(C)$, and isomorphic to D_4.

The proof of (iii) is in section 2 below. $\qquad\square$

Proposition 5. *Let C be a non empty bounded convex open subset in a space V of dimension n. Assume that the group $Coll(C)$ acts transitively on C.*

(i) If $n = 2$, then C is either the interior of an ellipse or an open triangle.

(ii) If $n = 3$, then C is the interior either of an ellipsoid, or of a cone over an ellipse, or of a tetrahedron.

(iii) If $n = 4$, there are six possible cases modulo the group $Coll(\overline{V}) \approx PGL_5(\mathbb{R})$. Each one has a representative C for which $\partial \overline{C}$ is contained in an algebraic hypersurface with equation

$$\xi_0 \xi_1 - \xi_2^2 - \xi_3^2 - \xi_4^2 = 0 \qquad \text{(\textit{C} is an open 4-ball)}$$
$$\xi_0 (\xi_1 \xi_2 - \xi_3^2 - \xi_4^2) = 0 \qquad \text{(cone based on a 3-ball)}$$
$$\xi_0 \xi_1 (\xi_2 \xi_3 - \xi_4^2) = 0 \qquad \text{(double cone on a 2-disc)}$$
$$\xi_0 \xi_1 \xi_2 \xi_3 \xi_4 = 0 \qquad \text{(4-simplex)}$$

$$(\xi_0\xi_1 - \xi_2^2)(\xi_0\xi_3 - \xi_4^2) = 0 \qquad \text{(intersection of two cones)}$$
$$\xi_0\xi_1\xi_2 - \xi_1\xi_3^2 - \xi_2\xi_4^2 = 0 \qquad \text{(convex hull of two ellipses whose}$$

planes meet in a single point common to the two ellipses)

where $[\xi_0 : \xi_1 : \xi_2 : \xi_3 : \xi_4]$ *are homogeneous coordinates on* $\overline{V}$.

In the statement of (iii) above, the word *cone* refers (say in the second case) to the interior C of the convex hull of $v \cup \Sigma$, where v is the vertex $[1 : 0 : 0 : 0 : 0] \in \overline{V}$ and where Σ is the 3-ball with boundary given by the equations

$$\xi_1\xi_2 - \xi_3^2 - \xi_4^2 = 0,$$
$$\xi_0 = 0.$$

The boundary $\partial\overline{C}$ lies in the union of two hypersurfaces, the hyperplane of equation $\xi_0 = 0$ and the quadric of equation $\xi_1\xi_2 - \xi_3^2 - \xi_4^2 = 0$. The latter is also called a *cone* by algebraic geometers, but the reader should be aware of two meanings for the same word.

Proposition 5 answers (for $n \leq 4$) a question formulated by H. Busemann in 1965 [Bu0]. In section 6 of [Bu2], under *extra smoothness* hypothesis on $\partial\overline{C}$, he showed that a homogeneous C is necessarily an ellipsoid (n arbitrary). Proposition 5 states part of the results obtained by Larman, Mani and Rogers [LMR]; as the latter is an unpublished manuscript, we have to rely on [Rog] which is an exposition of results (without proofs). I shall not discuss the case $n = 5$ (see [Rog]), but for mentioning the space *Sym* which appears in Remark 3.1 below. I shall no more discuss the relationship between homogeneous C's and various subjects (homogeneous cones and formally real Jordan algebras, works of M. Koecher, E.B. Vinberg and others) but for quoting the summary in chapter I of [Sat].

In connection with the remark following Proposition 3, I believe that $Isom(C)$ is transitive on C if and only if $Coll(C)$ is transitive on C.

2. The case of simplices

We consider first the two-dimensional case. Let V be an affine plane lying inside a projective completion $\overline{V}$ isomorphic to $\mathbb{P}^2(\mathbb{R})$, and let S_2 be the interior of a non degenerate triangle in V with vertices a, b, c. We view S_2 as a metric space supplied with its Hilbert metric. If S_2' is any other triangle in V with vertices a', b', c', there exists a projective linear isomorphism $g \in Coll(\overline{V})$ which maps S_2 (respectively a, b, c) onto S_2' (resp. a', b', c'), so that in particular S_2 and S_2' are isometric. Consequently, there is no loss of generality if we identify from now on V with the hyperplane of equation $\xi + \eta + \zeta = 1$ in $\mathbb{R}^3$ and if we consider the standard open simplex

$$S_2 = \left\{ \begin{pmatrix} \xi \\ \eta \\ \zeta \end{pmatrix} \in \mathbb{R}^3 \ : \ \xi > 0, \quad \eta > 0, \quad \zeta > 0, \quad \xi + \eta + \zeta = 1 \right\},$$

whose vertices

$$a = \begin{pmatrix} 1 \\ 0 \\ 0 \end{pmatrix}, \qquad b = \begin{pmatrix} 0 \\ 1 \\ 0 \end{pmatrix}, \qquad c = \begin{pmatrix} 0 \\ 0 \\ 1 \end{pmatrix}$$

constitute the canonical basis of $\mathbb{R}^3$.

We identify also the group $Coll(\overline{V})$ with the quotient $PGL_3(\mathbb{R})$ of $GL_3(\mathbb{R})$ by its center, the latter being the group $\mathbb{R}^*$ of homotheties. If $\begin{bmatrix} * & * & * \\ * & * & * \\ * & * & * \end{bmatrix}$ denotes the class in $PGL_3(\mathbb{R})$ of a matrix $\begin{pmatrix} * & * & * \\ * & * & * \\ * & * & * \end{pmatrix}$ in $GL_3(\mathbb{R})$, the subgroup

$$T_2 = \left\{ \begin{bmatrix} \lambda & 0 & 0 \\ 0 & \mu & 0 \\ 0 & 0 & \nu \end{bmatrix} \in PGL_3(\mathbb{R}) \quad : \quad \lambda > 0, \quad \mu > 0, \quad \nu > 0 \right\}$$

of $PGL_3(\mathbb{R})$ acts on S_2 by

$$\begin{bmatrix} \lambda & 0 & 0 \\ 0 & \mu & 0 \\ 0 & 0 & \nu \end{bmatrix} \begin{pmatrix} \xi \\ \eta \\ \zeta \end{pmatrix} = \frac{1}{\lambda \xi + \mu \eta + \nu \zeta} \begin{pmatrix} \lambda \xi \\ \mu \eta \\ \nu \zeta \end{pmatrix}.$$

The action is transitive because

$$\begin{bmatrix} \xi & 0 & 0 \\ 0 & \eta & 0 \\ 0 & 0 & \zeta \end{bmatrix} \begin{pmatrix} 1/3 \\ 1/3 \\ 1/3 \end{pmatrix} = \begin{pmatrix} \xi \\ \eta \\ \zeta \end{pmatrix},$$

for each $\begin{pmatrix} \xi \\ \eta \\ \zeta \end{pmatrix} \in S_2$. The images in $PGL_3(\mathbb{R})$ of the permutation matrices in $GL_3(\mathbb{R})$ constitute a symmetric group σ_3; this acts on S_2 and on $\overline{S}_2$, and permutes the vertices a, b, c. It is now straightforward to check that

$$Coll(S_2) = T_2 \rtimes \sigma_3,$$

where $Coll(S_2)$ is the group defined before Proposition 3 and where $\rtimes$ indicates the natural semi-direct product.

As T_2 acts transitively on S_2, spheres in S_2 are images by elements of T_2 of spheres around the base point $p_0 = \begin{pmatrix} 1/3 \\ 1/3 \\ 1/3 \end{pmatrix} \in S_2$. Let us describe one of these.

Consider a point $x_1 \in \ell_{a,p_0} \cap S_2$ distinct from p_0 and let $\Sigma_{x_1}(p_0)$ be the sphere in S_2 of center p_0 which contains x_1, as in Figure 5. Set $x_2 = \ell_{c,p_0} \cap \ell_{b,x_1}$. For any $y \in [x_1, x_2]$, one has $d(p_0, y) = d(p_0, x_1)$, because p_0, y and the two points of $\ell_{p_0,y} \cap \partial \overline{S}_2$ are on the four lines $\ell_{b,p_0}, \ell_{b,x_1}, \ell_{b,c}$ and $\ell_{b,a}$, which do not depend on y (we apply here once more the projective invariance of the cross-ratio as in Lemma 6.5.4 of [Ber]). Thus $[x_1, x_2] \subset \Sigma_{x_1}(p_0)$. Similar arguments show that this sphere is a hexagon with vertices $x_1, \ldots, x_6$ as in Figure 5. With the notations of the proof of Proposition 4, one has $\Sigma^!_{x_1}(p_0) = \{x_1, \ldots, x_6\}$.

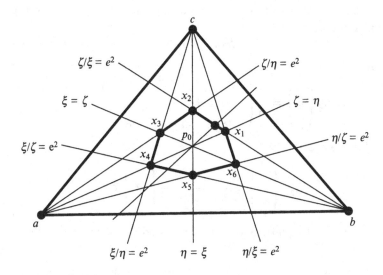

Figure 5: A sphere (!).

Denote by $r = d(p_0, x_1)$ the radius of $\Sigma_{x_1}(p_0)$. The line ℓ_{c,x_1} has the equation $\eta/\xi = e^r$. We have indicated in Figure 5 this equation, and more generally the equations of the lines going through one of the vertices of S_2 and one of the x_j's.

Observe that the isotropy subgroup of the base point $p_0 \in S_2$ in the group T_2 is reduced to $\{1\}$, so that the map

$$T_2 \to S_2, \qquad g \mapsto gp_0,$$

is a bijection (in other words, the action of T_2 on S_2 is *simply transitive*). Observe also that T_2 is naturally isomorphic to the subgroup

$$\tilde{T}_2 = \left\{ \begin{pmatrix} \lambda & 0 & 0 \\ 0 & \mu & 0 \\ 0 & 0 & \nu \end{pmatrix} \in GL_3(\mathbb{R}) \ : \ \lambda > 0, \quad \mu > 0, \quad \nu > 0, \quad \lambda\mu\nu = 1 \right\}$$

of $GL_3(\mathbb{R})$. Let us introduce, moreover, the vector space

$$W_2 = \left\{ \begin{pmatrix} \theta \\ \phi \\ \psi \end{pmatrix} \in \mathbb{R}^3 \; : \; \theta + \phi + \psi = 0 \right\}$$

and the group isomorphism $\widetilde{T}_2 \to W_2$ given by

$$\begin{pmatrix} \lambda & 0 & 0 \\ 0 & \mu & 0 \\ 0 & 0 & \nu \end{pmatrix} \mapsto \begin{pmatrix} \log \lambda \\ \log \mu \\ \log \nu \end{pmatrix}.$$

The appropriate composition of the maps introduced above and of their inverses is a bijection

$$\Lambda_2 : S_2 \to W_2,$$

given by

$$\Lambda_2 \begin{pmatrix} \xi \\ \eta \\ \zeta \end{pmatrix} = \frac{1}{3} \begin{pmatrix} 2\log \xi & - \log \eta & - \log \zeta \\ -\log \xi & + 2\log \eta & - \log \zeta \\ -\log \xi & - \log \eta & + 2\log \zeta \end{pmatrix} \in W_2,$$

and with inverse given by

$$\Lambda_2^{-1} \begin{pmatrix} \theta \\ \phi \\ \psi \end{pmatrix} = \frac{1}{e^\theta + e^\phi + e^\psi} \begin{pmatrix} e^\theta \\ e^\phi \\ e^\psi \end{pmatrix} \in S_2.$$

The lines in W_2 of equations $\phi = \psi$, $\psi = \theta$ and $\theta = \phi$ are called the *axis* of W_2.

Proposition 6. *The bijection $\Lambda_2 : S_2 \to W_2$ defined above has the following properties.*

(i) *Lines in S_2 pointing to one of a, b, c are mapped by Λ_2 onto lines in W_2 parallel to one of the axis.*

(ii) *For each real number $r > 0$, the lines in S_2 of equations $\xi/\eta = e^{\pm r}$ are mapped by Λ_2 onto the lines in W_2 of equations $\theta - \phi = \pm r$. Similarly for $\eta/\zeta = e^{\pm r}$ and $\phi - \psi = \pm r$, as well as for $\zeta/\xi = e^{\pm r}$ and $\psi - \theta = \pm r$.*

(iii) *Spheres in S_2 are mapped by Λ_2 onto regular hexagons in W_2 with sides parallel to the axis.*

(iv) *Straight lines in S_2 through p_0 are in general not mapped by Λ_2 onto straight lines in W_2.*

Proof. This is straightforward, from the formulas for Λ_2 and Λ_2^{-1}. By "regular hexagon" in W_2, we mean the image by some translation of a hexagon with the two following properties: firstly its three main diagonals intersect in the origin $O \in W_2$, and secondly any of its vertices is the vector sum of the two nearest vertices.   □

Digression on normed spaces

Let U be a real vector space, say finite dimensional here for simplicity. Let B be an open neighbourhood of the origin in U which is bounded, convex and symmetric $(x \in U \iff -x \in U)$. Define a function $p_B : U \to [0, \infty[$ by

$$p_B(x) = \inf \{\rho \in \mathbb{R} \ : \ \rho > 0 \text{ and } x \in \rho B\}$$

(p_B is the so-called *Minkowski functional* for B — see e.g. [KeN]).

Lemma. *The function p_B is a norm on the vector space U.*

Proof. Consider first $\rho, \sigma > 0$ and $u, v \in B$. Set

$$w = \frac{\rho}{\rho + \sigma} u + \frac{\sigma}{\rho + \sigma} v.$$

Then $w \in B$ by convexity, and $\rho u + \sigma v = (\rho + \sigma) w \in (\rho + \sigma)B$. Thus $\rho B + \sigma B = (\rho + \sigma)B$.

Consider then $x, y \in U$. If $\rho, \sigma > 0$ are such that $x \in \rho B$ and $y \in \sigma B$, one has

$$x + y \in \rho B + \sigma B = (\rho + \sigma)B.$$

It follows that $p_B(x + y) \le p_B(x) + p_B(y)$.

The other properties entering the definition of a norm $(p_B(\lambda x) = |\lambda| p_B(x)$, and $p_B(x) = 0 \iff x = 0$) are straightforward to check. $\square$

In case U is a plane and B is a regular hexagon, we say that the resulting normed space in a *hexagonal plane*.

As M.A. Roller has observed, there is an alternative way to describe the metric in this space: start with a plane U given together with a Euclidean metric and a hexagon B which is regular in the Euclidean sense (edges of equal lengths and equal angles); then the hexagonal distance between two points $x, y \in U$ is the shortest Euclidean length of a polygonal line from x to y with each of its sides parallel to some edge of B. (Similarly, if B was a square, one would obtain the so-called Manhattan metric on U. All regular polygons with an even number of vertices provide variations on this theme.) This ends the digression on normed spaces.

Let again $\Lambda_2 : S_2 \to W_2$ be as in Proposition 6. Denote by B the regular hexagon in W_2 containing 0 and having its six edges on the lines of equations

$$|\theta - \phi| = 1, \qquad |\phi - \psi| = 1, \qquad |\psi - \theta| = 1.$$

(Observe that these edges are perpendicular to the positive roots

$$\alpha_1 = \begin{pmatrix} 1 \\ -1 \\ 0 \end{pmatrix}, \qquad \alpha_2 = \begin{pmatrix} 0 \\ 1 \\ -1 \end{pmatrix}, \qquad \alpha_1 + \alpha_2 = \begin{pmatrix} 1 \\ 0 \\ -1 \end{pmatrix}$$

of the usual root system of type A_2 in W_2, and compare with the end of the present section.) Consider W_2 as a hexagonal plane for the norm having B as its unit ball. The following result is both a part of and a complement to Theorem 1 of [Woj].

Proposition 7. *The map $\Lambda_2 : S_2 \to W_2$ is an isometry from the 2-simplex S_2 (with Hilbert's metric) onto the hexagonal plane W_2.*

Proof. Let $x', x'' \in S_2$ be such that $x' \neq x''$. Assume first that the line $\ell_{x',x''}$ contains a vertex of S_2, say c, and let $\xi/\eta = e^r$ be the equation of $\ell_{x',x''}$. One has

$$x' = \begin{pmatrix} \xi' \\ \eta' \\ \zeta' \end{pmatrix} \in S_2, \qquad \xi' = e^r \eta', \qquad \xi' + \eta' + \zeta' = 1,$$

$$x'' = \begin{pmatrix} \xi'' \\ \eta'' \\ \zeta'' \end{pmatrix} \in S_2, \qquad \xi'' = e^r \eta'', \qquad \xi'' + \eta'' + \zeta'' = 1,$$

$$d(x', x'') = \log \left(\frac{\zeta'/\eta'}{\zeta''/\eta''} \right).$$

Set $r' = \log(\zeta'/\eta')$ and $r'' = \log(\zeta''/\eta'')$, so that $d_S(x', x'') = r' - r''$ (where d_S denotes Hilbert's metric inside S_2 and where notations are such that $r' > r''$).

If $\Lambda_2(x') = \begin{pmatrix} \theta' \\ \phi' \\ \psi' \end{pmatrix}$ one has

$$\phi' - \theta' = -r, \qquad \psi' - \phi' = r', \qquad \theta' + \phi' + \psi' = 0;$$

thus

$$\Lambda_2(x') = \frac{1}{3} \begin{pmatrix} 2r - r' \\ -r - r' \\ -r + 2r' \end{pmatrix},$$

and similarly for $\Lambda_2(x'')$, so that

$$\Lambda_2(x'') - \Lambda_2(x') = (r' - r'') \begin{pmatrix} 1/3 \\ 1/3 \\ -2/3 \end{pmatrix} \in W_2.$$

As $\begin{pmatrix} 1/3 \\ 1/3 \\ -2/3 \end{pmatrix}$ is in the boundary of the hexagon B, one has

$$d_W(\Lambda_2(x'), \Lambda_2(x'')) = ||\Lambda_2(x'') - \Lambda_2(x')|| = r' - r'' = d_S(x', x''),$$

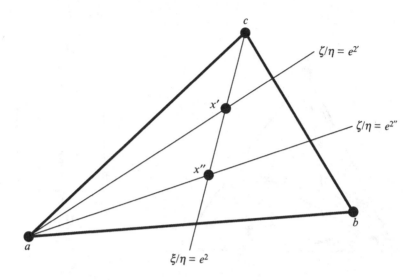

Figure 6: First step in proof of Proposition 7.

(where d_W denotes the hexagonal metric on W_2) as in Figure 6.
Consider now the generic case of two distinct points $x', x'' \in S_2$ such that $\ell_{x',x''}$ does not contain any vertex of S_2. Assume the notations such that

$$d(x'', \ell_{b,x'}) \leq \min\{d(x'', \ell_{a,x'}), d(x'', \ell_{c,x'})\}.$$

Choose $p \in \ell_{b,x'}$ such that x', x'' are in a sphere $\Sigma(p)$ of S_2 with center p and corners $x_1 = x'$, $x_2, \ldots, x_6$ in cyclic order. Using the genericity assumption and the previous inequality one may check that $x'' \in]x_3, x_4[\cup]x_4, x_5[$.
The image $\Lambda_2(\Sigma(p))$ is a regular hexagon in W_2 that one may obtain from B by translation and homothety. One checks easily that

$$d(x', x'') = d(x', x_4),$$

and that

$$\|\Lambda_2(x'') - \Lambda_2(x')\| = \|\Lambda_2(x_4) - \Lambda_2(x')\|.$$

As $d(x', x_4) = \|\Lambda_2(x_4) - \Lambda_2(x')\|$ by the first part of the proof, one has indeed

$$d(x', x'') = \|\Lambda_2(x'') - \Lambda_2(x')\|,$$

and the proof is complete. (I am most grateful to Pierre Planche for correcting an earlier incorrect proof of this proposition). $\quad\square$

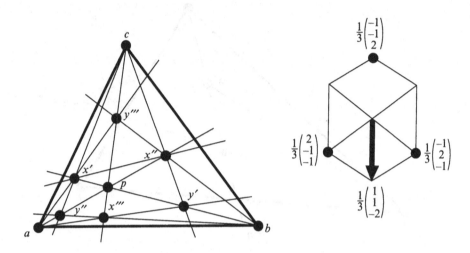

Figure 7: Second step in proof of Proposition 7.

Corollary. *The full isometry group of S_2 is isomorphic to the natural semi-direct product $\mathbb{R}^2 \rtimes D_6$, where D_6 denotes the dihedral group of order 12.*

Proof. This follows from Proposition 7 because an isometry of a real normed space is necessarily an affine map [MaU]. □

Proposition 7 shows in particular that S_2 is *quasi-isometric* to a Euclidean plane, because two normed real vector spaces of equal finite dimensions are quasi-isometric.

It implies also the following. Let Γ denote the isometry $x \mapsto -x$ of W_2, and let $\gamma = \Lambda_2^{-1}\Gamma\Lambda_2$ be the corresponding isometry of S_2.

Proposition 8. *The isometry γ of S_2 fixes the base point $p_0 = \begin{pmatrix} 1/3 \\ 1/3 \\ 1/3 \end{pmatrix}$. If*

ℓ is a line in S_2 through p_0, then $\gamma(\ell)$ is a line if and only if ℓ contains one of the vertices a, b, c of $\overline{S}_2$. In formulas, γ is the quadratic transformation

$$\gamma : \begin{pmatrix} \xi \\ \eta \\ \zeta \end{pmatrix} \mapsto \frac{1}{1/\xi + 1/\eta + 1/\zeta} \begin{pmatrix} 1/\xi \\ 1/\eta \\ 1/\zeta \end{pmatrix}$$

of S_2, which blows up the vertices a, b, c of $\overline{S}_2$ and blows down the edges $]b,c[$, $]c,a[$, $]a,b[$.

In particular γ is not in Coll(S_2) *and does* not *extend to a continuous transformation of the closed 2-simplex* $\overline{S}_2$.

For *quadratic transformations* and the related Cremona's transformations, see for example numbers IV.11 and VI.56 in [Die].

We consider now briefly the case of the n-simplex

$$S_n = \{(\xi_i)_{1 \le i \le n+1} \in \mathbb{R}^{n+1} \; : \; \xi_i > 0 \text{ and } \xi_1 + \cdots + \xi_{n+1} = 1\}.$$

Set

$$W_n = \{(\theta_i)_{1 \le i \le n+1} \in \mathbb{R}^{n+1} \; : \; \theta_1 + \cdots + \theta_{n+1} = 0\},$$

and define $\Lambda_n : S_n \to W_n$ by

$$\Lambda_n \begin{pmatrix} \xi_1 \\ \vdots \\ \xi_{n+1} \end{pmatrix} = \begin{pmatrix} \theta_1 \\ \vdots \\ \theta_{n+1} \end{pmatrix}, \qquad \text{where} \quad \theta_i = \frac{1}{n+1}\left(n \log \xi_i - \sum_{j \ne i} \log \xi_j\right).$$

The trace on S_n of a linear hyperplane in $\mathbb{R}^{n+1}$ containing a $(n-1)$-face of $\overline{S}_n$ is mapped by Λ_n onto a hyperplane of W_n with an equation of the form $\theta_i - \theta_j = const$ for some pair (i,j) of distinct indices in $\{1,\ldots,n+1\}$. Let $(\epsilon_i)_{1 \le i \le n+1}$ denote the canonical basis of $\mathbb{R}^{n+1}$. Consider in W_n the usual root system of type A_n, consisting of the $n(n+1)$ vectors $\epsilon_i - \epsilon_j$, for $i, j \in \{1,\ldots,n+1\}$ and $i \ne j$ (see e.g. [Bou]). The images by Λ_n of the appropriate unit ball of S_n is the symmetric convex set

$$B_n = \{(\theta_i)_{1 \le i \le n+1} \in W_n \; : \; |\theta_i - \theta_j| \le 1 \text{ for } i \ne j \in \{1,\ldots,n+1\}\} \subset W_n.$$

One may show again that Λ_n is an isometry from S_n onto the space W_n furnished with a norm for which B_n is the unit ball.

If $n = 3$, one replaces W_3 by the usual space $\mathbb{R}^3$, with coordinates $\begin{pmatrix} \theta \\ \phi \\ \psi \end{pmatrix}$, and B_3 by the convex hull of 14 points which are

$$\begin{pmatrix} \pm 2 \\ 0 \\ 0 \end{pmatrix}, \quad \begin{pmatrix} 0 \\ \pm 2 \\ 0 \end{pmatrix}, \quad \begin{pmatrix} 0 \\ 0 \\ \pm 2 \end{pmatrix},$$

(6 vertices of degree 4 in the 1-skeleton of B_3) and

$$\begin{pmatrix} \pm 1 \\ \pm 1 \\ \pm 1 \end{pmatrix}$$

(8 vertices of degree 3). The faces of B_3 are then quadrilaterals lying in the
12 planes of equations

$$|\theta \pm \phi| = 2, \qquad |\phi \pm \psi| = 2, \qquad |\psi \pm \theta| = 2.$$

The polytope B_3 has consequently 14 vertices, 24 edges and 12 faces. (It is
a so-called rhombic dodecahedron, that crystallographs know as the Voronoi
polyhedron of a face-centered cubic lattice; see §18.3 and §22.4 in [Cox].)
The 12 faces of B_3 are in affine planes which are perpendicular (with respect
to the canonical Euclidean structure on $\mathbb{R}^3$) to the roots of the root system
$R = R_+ \sqcup (-R_+)$ of type A_3, where R_+ consists of the six positive roots

$$\alpha_1 = \begin{pmatrix} 0 \\ 2 \\ 2 \end{pmatrix}, \qquad \alpha_2 = \begin{pmatrix} 2 \\ -2 \\ 0 \end{pmatrix}, \qquad \alpha_3 = \begin{pmatrix} 0 \\ 2 \\ -2 \end{pmatrix},$$

$$\alpha_1 + \alpha_2 = \begin{pmatrix} 2 \\ 0 \\ 2 \end{pmatrix}, \qquad \alpha_1 + \alpha_2 + \alpha_3 = \begin{pmatrix} 2 \\ 2 \\ 0 \end{pmatrix}, \qquad \alpha_2 + \alpha_3 = \begin{pmatrix} 2 \\ 0 \\ -2 \end{pmatrix}.$$

More precisely:

    the planes of equations $|\phi - \psi| = 2$ are perpendicular to α_1,
    the planes of equations $|\theta + \phi| = 2$ are perpendicular to α_2,
    the planes of equations $|\phi + \psi| = 2$ are perpendicular to α_3,
    the planes of equations $|\theta - \psi| = 2$ are perpendicular to $\alpha_1 + \alpha_2$,
    the planes of equations $|\theta - \phi| = 2$ are perpendicular to $\alpha_1 + \alpha_2 + \alpha_3$,
    the planes of equations $|\theta + \psi| = 2$ are perpendicular to $\alpha_2 + \alpha_3$.

The relation between the polytope B_n and the usual root system of type A_n
in W_n is similar.

3. Remarks and questions

3.1. Let *Sym* denote the 5-dimensional real affine space of real symmetric
3-by-3 matrices of trace 1. Consider the subset of definite positive matrices

$$Pos = \left\{ x = \begin{pmatrix} \xi & \tau & \sigma \\ \tau & \eta & \rho \\ \sigma & \rho & \zeta \end{pmatrix} \in Sym \;:\; \xi > 0, \; \begin{vmatrix} \xi & \tau \\ \tau & \eta \end{vmatrix} > 0, \; \det(x) > 0 \right\}.$$

Then *Pos* is obviously open and convex in *Sym*. Moreover *Pos* is bounded;
indeed, with the notations above, one has for any $g \in Pos$

$$\xi > 0, \qquad \eta > 0, \qquad \zeta > 0,$$

so that $\xi, \eta, \zeta \in]0, 1[$ (because $\xi + \eta + \zeta = 1$), and

$$\begin{vmatrix} \eta & \rho \\ \rho & \zeta \end{vmatrix} > 0, \qquad \begin{vmatrix} \xi & \sigma \\ \sigma & \zeta \end{vmatrix} > 0, \qquad \begin{vmatrix} \xi & \tau \\ \tau & \eta \end{vmatrix} > 0,$$

so that $\rho, \sigma, \tau \in] -1, 1[$. (For the criterion for positive definiteness in terms of minors, see e.g. §X.4 in [Gan].) From now on, we consider *Pos* as a metric space with the Hilbert metric.

The group $G = SL_3(\mathbb{R})$ acts on *Pos* by

$$G \times Pos \longrightarrow Pos, \qquad (g, x) \mapsto \frac{1}{trace(gxg^t)} gxg^t, \qquad (*)$$

where g^t denotes the transpose of a matrix g. Standard facts about the diagonalization of symmetric matrices show that this action is *transitive*, and that the isotropy subgroup in G of the pase point

$$x_0 = \frac{1}{3} \begin{pmatrix} 1 & 0 & 0 \\ 0 & 1 & 0 \\ 0 & 0 & 1 \end{pmatrix} \in Pos$$

is the orthogonal group

$$K = SO(3) = \{g \in G : gg^t = 1\}.$$

Consequently, *Pos* is a homogeneous space

$$Pos = G/K,$$

which can be taken as the simplest case of a non compact Riemannian symmetric space of real rank > 1 . (More generally $Pos_{n+1} = SL_{n+1}(\mathbb{R})/SO(n+1)$ is a Riemannian symmetric space of real rank n ; the fact that positive matrices $x \in Pos_{n+1}$ are normalized here by $trace(x) = 1$ and in most other places by $det(x) = 1$ is not important.)

The closure $\overline{Pos}$ of *Pos* in *Sym* is the space of (not necessarily definite) positive 3-by-3 matrices of trace 1. Thus formula $(*)$ above defines also an action of G on $\overline{Pos}$ which preserves cross-ratios. It follows that G is a *transitive group of isometries for Hilbert's metric*. Of course, G is also a transitive group of isometries for the *Riemannian metric* (see e.g. [Hel] or [Mos]) which is more usual in this setting. But the following considerations indicate that it should be rewarding to study seriously Hilbert's metric on *Pos*.

Let us identify the open simplex S_2 of section 2 with the subset of *Pos* consisting of diagonal matrices. The restriction to S_2 of the Hilbert metric on *Pos* coincides with the Hilbert metric on S_2 as studied in section 2. The *flats* of *Pos* are precisely the images of S_2 by the elements of G. One should be able to caracterize the flats among all ideal 2-simplices of *Pos*, using Hilbert's metric on *Pos* only, and *not* the action of *G*. Now it is a fact that a generic Riemannian geodesic of *Pos* in S_2 through the base point x_0 belongs to a unique flat (which is S_2!), and that there are exactly three non generic exceptions, defining 6 chambers in the flat S_2; these are the chambers entering the definition of the Furstenberg boundary of *Sym*. In this way the fine structure of the Riemannian geodesics is rather subtle. One may hope that Hilbert's geometry will suggest another way to look at geodesics and another model of the Furstenberg boundary which will prove useful.

In the study of real hyperbolic spaces, the notion of convexity and the various kinds of simplices (compact, ideal, regular) play a fundamental role: see e.g. Gromov's proof of Mostow's rigidity theorem [Mun]. Whereas the classical Riemannian approach to symmetric spaces makes it difficult to guess how convexity arguments could be generalized to spaces of real rank ≥ 2, the Hilbert approach sketched above provides (here for *Pos*) obvious candidates for convex subsets in general and for simplices in particular. This suggests immediately a systematic investigation of simplices in *Pos*. As a first step, one could for example try to classify the isometry types of the 2-dimensional sections of *Pos* (and more generally of Pos_{n+1}).
This programme is due to N. A'Campo.

3.2. There are various open questions about Hilbert's metric, even in low dimensions. Let us mention the following CAT(0)-problem.
Let S_2 be the interior of a triangle with vertices a, b, c in $\mathbb{R}^2$, let w, x, y be the three vertices of a triangle inside S_2 and let z be a point on $[x,y]$. If the configuration is as in Figure 8, it is obvious that $d(w,z) = d(w,x) = d(w,y)$. (The configuration has been chosen in the Figure in such a way that there is a *unique* geodesic segment inside S_2 joining each of the pairs (w,x), (x,y), (y,w).) Thus S_2 is far from satisfying the CAT(0)-criterion. (We refer to [Tro] for an exposition of CAT(0), which holds for Comparison, Alexandrov, Toponogov and 0-curvature.) Similarly, if C is a neighborhood of S_2 such that $\overline{C}$ is strictly convex, as in Figure 8, the space C need not satisfy CAT(0). On the other hand, Hilbert's metric on the interior of an ellipse does satisfy CAT(0), because it is isometric to the Poincaré disc. It would be interesting to know which are the bounded convex open sets which satisfy CAT(0).

3.3. The definition of Hilbert's metric on a convex subset C of a real vector space V makes sense even if V is *infinite dimensional*. The openess and boundedness conditions on C mean that the intersection of any affine line ℓ with C has to be an open and bounded segment in ℓ. For example, consider the Hilbert space

$$\ell_{\mathbb{R}}^2(\mathbb{N}) = \{(x_n)_{n \geq 1} : x_n \in \mathbb{R} \text{ and } \sum |x_n|^2 < \infty\}$$

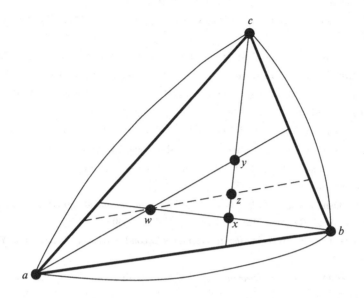

Figure 8: On CAT(0).

and the subset

$$C = \{(x_n)_{n\geq 1} \in \ell^2_{\mathbb{R}}(\mathbb{N}) \ : \ |x_n| < n! \ \text{for all} \ n > 1\}.$$

Then C satisfies the conditions above, though C does not lie in any bounded ball of $\ell^2_{\mathbb{R}}(\mathbb{N})$ in the usual sense. It could be interesting to investigate Hilbert's metric on such a C.

Of course, there are much more natural examples of C's which deserve further study. An obvious candidate is the open unit ball $A_{<1}$ of a C*-algebra A (say with unit). To end with a brave speculation, one could suggest that Hilbert's metric on $A_{<1}$ has something to do with the offspring of the Russo-Dye Theorem [Gar], [KaP].

Acknowledgements. Much of the above was shown to me a few years ago by Norbert A'Campo. Since then, it has been the subject of several leisurely discussions involving among others Albert Fathy, Etienne Ghys and Raffaele Russo. I am very grateful to them.

References

[Bea]   H.S. Bear, *Part Metric and Hyperbolic Metric*, Am. Math. Monthly **98** (1991), pp. 109–123.

[Ber]   M. Berger, *Geometry I, II*, Springer, New York, 1987.

[Bir]   G. Birkhoff, *Extensions of Jentzsch's Theorem*, Trans. Amer. Math. Soc. **85** (1957), pp. 219–227.

[Bou]   N. Bourbaki, *Groupes et algèbres de Lie*, chapitres 4, 5, 6, Hermann, Paris, 1968.

[Brø]   A. Brøndsted, *An Introduction to Convex Polytopes*, Springer, New York, 1983.

[Bu0]   H. Busemann, *Problem 3*, in: Proc. of the Colloquium on Convexity (W. Fenchel, ed.), Copenhagen, 1965, p. 309.

[Bu1]   H. Busemann, *The Geometry of Geodesics*, Academic Press, New York, 1955.

[Bu2]   H. Busemann, *Timelike Spaces*, Dissert. Math. vol. 53, Warsaw, 1967.

[BuK]   H. Busemann and P.J. Kelly, *Projective Geometry and Projective Metrics*, Academic Press, New York, 1953.

[Bus]   P.J. Bushell, *Hilbert's Metric and Positive Contraction Mappings in a Banach Space*, Arch. Rational Mech. Anal. **52** (1973), pp. 330–338.

[Cox]   H.S.M. Coxeter, *Introduction to Geometry*, second edition, J. Wiley, New York, 1969.

[Die]   J. Dieudonné, *Cours de géométrie algébrique 1*, Presses Univ. de France, Paris, 1974.

[Gan]   F.R. Gantmacher, *The Theory of Matrices*, Vol. 1, Chelsea, New York, 1959.

[Gar]   L.T. Gardner, *An Elementary Proof of the Russo-Dye Theorem*, Proc. Amer. Math. Soc. **90** (1984), p. 171.

[Hel]   S. Helgason, *Differential Geometry and Symmetric Spaces*, Academic Press, New York, 1962.

[Hil]   D. Hilbert, *Ueber die gerade Linie als kürzeste Verbindung zweier Punkte*, Math. Ann. **46** (1895), pp. 91–96.

[KaP]   R.V. Kadison and G.K. Pedersen, *Means and Convex Combinations of Unitary Operators*, Math. Scand. **57** (1985), pp. 249–266.

[KeN]   J.L. Kelley and I. Namioka, *Linear Topological Spaces*, Van Nostrand, Princeton, 1963.

[KoP]   E. Kohlberg and J.W. Pratt, *The Contraction Mapping Approach to the Perron-Frobenius Theory: why Hilbert's Metric?*, Math. Operations Research **7** (1982), pp. 198–210.

[LMR]   D.G. Larman, P. Mani and C.A. Rogers, *The Transitive Action of a Group of Projectivities on the Interior of a Convex Body*, unpublished manuscript, announced in the following reference.

[LMR'] C.A. Rogers, *Convex Bodies That Are Invariant Under a Group of Projectivities That Act Transitively on Their Interiors*, Bull. London Math. Soc. **8** (1976), p. 20.

[MaU]   S. Mazur and S. Ulam, *Sur les transformations isométriques d'espaces vectoriels normés*, Comptes Rendus **194** (1932), pp. 946–948.

[Mos]   D. Mostow, *Some New Decomposition Theorems for Semi-simple Groups*, Mem. Amer. Math. Soc. **14** (1955), pp. 31–54.

[Mun] H.J. Munkholm, *Simplices of Maximal Volume in Hyperbolic Space, Gromov's Norm, and Gromov's Proof of Mostow's Rigidity Theorem (Following Thurston)*, Lecture Notes in Math. 788, Springer, 1980, pp. 109–124.

[Rog] C.A. Rogers, *Some Problems in the Geometry of Convex Bodies*, in: The geometric vein (the Coxeter Festschrift), (C. Davis, B. Grünbaum and F.A. Sherk, eds.), Springer, 1981, pp. 279–284.

[Sae] H. Samelson, *On the Perron-Frobenius Theorem*, Michigan Math. J. 4 (1956), pp. 57–59.

[Sam] P. Samuel, *Géométrie projective*, Presses Universitaires de France, Paris, 1986.

[Sat] I. Satake, *Algebraic Structures of Symmetric Domains*, Iwanami Shoten and Princeton Univ. Press, 1980.

[Tro] M. Troyanov, *Espaces à courbure négative et groupes hyperboliques*, in: Sur les groupes hyperboliques d'après Mikhael Gromov (E. Ghys and P. de la Harpe, eds.), Birkhäuser, 1990, pp. 47–66.

[Woj] M. Wojtkowski, *On Uniform Contraction Generated by Positive Matrices*, in: Random matrices and their applications (J.E. Cohen et al., eds.), Contemp. Math. 50 (1986), pp. 109–118.

Software for Automatic Groups, Isomorphism Testing and Finitely Presented Groups

Derek F. Holt and Sarah Rees

Mathematics Institute, University of Warwick, Coventry CV4 7AL.
Department of Mathematics and Statistics, Merz Court, The University,
Newcastle-upon-Tyne NE1 7RU.

Introduction

The purpose of this note is to give a brief description of some software that has been developed by the authors (partly in collaboration with David Epstein), and which was demonstrated by them at the Workshop at the University of Sussex. It is all written in the C language and designed for use on UNIX systems. It is available for distribution (via ftp or SUN cartridge) free of charge from either of the authors, with source code and documentation included, mainly in the form of UNIX-style manual entries. Input and output is all done using files in a specially designed format. Usually, the user only needs to create files containing group presentations prior to running the programs. There are three principal components of this package, which we shall describe individually.

1. Automata

This is a sequence of programs that is designed to compute the automatic structure of a short-lex automatic group. For general information on automatic groups see [ECHLPT 92], and for a detailed description of the algorithms employed in these programs see [EHR 91]. These programs were written together with David Epstein.

Let $G = \langle X \mid R \rangle$ be a finitely presented group, where X is ordered and closed under inversion. For each $g \in G$, let $w(g)$ be the lexicographically least (using the given ordering of X) amongst the shortest words in X that represent g. Then G is said to be *short-lex automatic* (with respect to the ordered set X) if the following hold.

(i) There is a finite state automaton W called the *word-acceptor* with input alphabet X, which accepts precisely the words $w(g)$ for $g \in G$.

(ii) For each $x \in X$, there is a finite state automata $M(x)$ with input alphabet $(X \cup \{\$\}) \times (X \cup \{\$\})$, which accepts the pair of words (w_1, w_2) if and only if w_1 and w_2 are both accepted by W and $w_1 x$ and w_2 represent the same element of G. (The extra symbol $\$$ is used only to append on to the end of the shorter of the two words w_1 and w_2 to make them have the same lengths.) The automata $M(x)$ are known as the *multipliers*.

It is important to note that, although the general property of a group being automatic is independent of the generating set, the property of being short-lex automatic may depend on both the choice of X and on its ordering.

The input to the programs as supplied by the user is simply a file containing a finite group presentation, where the ordering of the generators is defined by the order in which they appear. In principal, if the group is short-lex automatic with these generators, then, given enough time and space, the programs will compute the word-acceptor and multipliers, and verify that they are correct. These are all output to files in our standard format. For examples requiring more time or space than is available, it can happen that a word-acceptor is produced, or even a word-acceptor and multipliers, but without a complete verification. In this case, it is possible but not certain that they will be correct. In any case W will accept all of the words $w(g)$, but in case of incorrectness it may accept some additional words.

Some other programs are available that operate on these automata. For example it is possible to enumerate systematically the accepted words, to decide whether this set is finite or infinite, and to determine its cardinality if it is finite. The finiteness and the order of G can therefore be determined, provided that the verification process has been completed. There is also a program that will draw a picture of a finite state automaton, as a labelled graph, in an X-window; the positions of the points and labels can be adjusted interactively, and a postscript version of the final result can be generated. Finally, there is a program (written by Uri Zwick in C++) that can compute the growth-rate polynomial of the group from the word-acceptor.

Currently, the automata package can complete within a reasonable period of time on various classes of examples. These include Euclidean groups, Coxeter groups with up to 4 generators, 2-dimensional surface groups, some 3-dimensional hyperbolic groups, and some knot groups. We are constantly attempting to extend its scope to include more examples.

2. Isomorphism testing

This is a program that attempts to decide whether two given finitely presented groups are isomorphic or not. A more detailed description can be found in [HoR 92a]. This problem is of course known to be undecidable in general, and so any such program is bound to fail on some inputs, but we might nevertheless hope to achieve success in many cases. The basic idea is to try to prove isomorphism and non-isomorphism alternately, for increasing periods of time, hoping to succeed eventually in one of these two aims. It is also possible, of course, to instruct the program to look only for an isomorphism, or for a proof of non-isomorphism. The package is arranged as a master program which calls other programs to perform specific calculations on the groups. These other programs can also be used individually if required. This can be advantageous if the user has some idea in advance of which particular calculations are likely to be the most helpful or decisive. The user has only to create files containing the two group presentations before starting.

We attempt to prove isomorphism, by running the Knuth-Bendix procedure on the group presentations (see, for example, [Gil 79]), in order to generate a word reduction algorithm for words in the generators. In cases of success, this will enable us to verify that a particular map from one group to the other is an isomorphism. In order to find this map, we have to use an exhaustive search, which often becomes rapidly impractical as the word lengths of the images of the generators under the map increase. We can offset this a little by using various additional tests, such as checking that the induced map between the abelian quotients is an isomorphism, or capable of being extended to an isomorphism when only some of the generator images are known. In addition, if both groups can be mapped onto a suitable finite permutation group, then we can check that our map or partial map can induce an automorphism of the permutation group. In principle, if the two groups are isomorphic, then this algorithm would eventually find an explicit isomorphism, but in practice the required time would eventually be prohibitive in many cases.

We attempt to prove non-isomorphism by examining the finite (and abelian-by-finite) quotients of the two groups, and checking to see if they correspond. This method was used successfully by Havas and Kovács to distinguish between various knot groups (see [HaK 84]). Unlike the isomorphism test, this approach cannot be guaranteed to work in general, even in principle, since there are certainly non-isomorphic groups whose finite quotients are the same; for example, finitely presented infinite simple groups. Furthermore, for reasons of space, it is only possible to look for finite quotients of reasonably small order, and in the case of finite nonabelian simple images, we have to look for epimorphisms onto each simple group individually. However, for various classes of groups that arise naturally from geometry and topology, such as knot groups and fundamental groups of manifolds, this approach seems to be quite successful. Indeed, this is one of the applications that we had in mind when planning this program. More details of the method for searching for finite quotients are given in the next section.

This program has been applied successfully to a number of genuine problems that have been passed on to us. For example, in [Kel 90], a collection of about 30 pairs of link groups is considered, each pair of which came from identical links with different orientations. All of these are groups with four generators and up to four relations. Some of these pairs were isomorphic and some not, and the program was able to decide this in all cases. Two of these cases took some hours of cpu-time, but all of the others were very quick.

3. Quotpic

This is a graphics package that the authors are currently developing for plotting the finite quotient groups of a given finitely presented group G. A more detailed description will appear in [HoR 92b]. For the graphics display, an X-windows server is required, which may be coloured or black and white. The user must first make a file containing the given presentation of G; thereafter, all computations are done interactively, by using a mouse to select menu options. A finite quotient $Q = G/N$ of G is represented in the display by a vertex, which can be thought of as being labelled by the normal subgroup N. Given two such vertices M and N such that $|G/M|$ has more prime factors (counted with repetition) than $|G/N|$, M will always appear lower than N. If N is joined to M either by an edge, or by a sequence of edges passing through intermediate vertices, then the inclusion $M \subset N$ holds. Initially, there will be a single vertex G corresponding to the trivial quotient; further vertices are plotted as the corresponding quotients are calculated. The system can output a postscript file for producing a good hard copy of the lattice that has been plotted. It operates by calling up other programs to perform individual calculations; these are more or less the same programs that are used in the isomorphism testing package for computing finite quotients.

Since all groups G of moderately small order are extensions of a soluble group N by a group T, where T is either trivial or satisfies $S \subseteq T \subseteq \text{Aut}(S)$ with S a direct product of isomorphic nonabelian simple groups, we have provided the system with two basic facilities. The first is to find the epimorphisms from G onto the relevant subgroups of $\text{Aut}(S)$, for all appropriate groups S. This is done interactively by the user, by selecting the group S from a list on the menu. (The list of groups S can easily be extended if required.) The second facility can be described as follows. Given a finite quotient G/M of G of sufficiently small order (a few thousand), find those quotients G/N of G with $N \subset M$ and M/N an elementary abelian p-group for some prime p. We achieve this as follows. First, we use an abelianized Reidemeister-Schreier process to compute the largest elementary abelian p-quotient $M/M^{(p)}$ of M for a particular prime p (which the user can select from the menu). (This process was first implemented by Havas in [Hav 74].) Then $M/M^{(p)}$ can be regarded as a $K(G/M)$-module, where K is the finite field of order p, and the normal subgroups N that we are seeking correspond precisely to the $K(G/M)$-submodules $N/M^{(p)}$ of $M/M^{(p)}$. In fact there was already in existence an efficient collection of programs for computing the submodule lattice of a module over a prime field, based on the well-known 'Meataxe' algorithm of Richard Parker (see [Par 84] and [LMR 92]). This had been written at Aachen, and we were kindly given permission to incorporate these programs into our system.

In addition to these basic facilities, we have provided a number of other options, which experience has suggested are useful in examples which arise in practice. For example, when G/M is not too large, we can compute the complete set of abelian invariants of $M/[M,M]$, and thereby determine which prime numbers p are relevant. We have also provided a version of the Canberra p-quotient algorithm (often known as the NQA; see, for example, [HaN 80]) to compute the quotients of the lower p-central series of M. The basic lattice operations intersection and join on the vertices (thought of as normal subgroups of G) are also available; indeed, these operations are used to ascertain whether or not one vertex is contained in another when this is not already clear. Although we are not providing sufficient facilities to completely identify any given quotient up to isomorphism, this is possible in many cases. As an easy aid to this, it is possible to count the numbers of elements of each order in a quotient. This is particularly useful if we are trying to distinguish one group from another. The simplest example is the use of this technique to distinguish between the dihedral and quaternion groups of order 8.

References

[B-M]   M. Bestvina and G. Mess, *The boundary of negatively curved groups*, to appear in J. Amer. Math. Soc.

[ECHLPT 92] D.B.A. Epstein, J.W. Cannon, D.F. Holt, S. Levy, M.S. Paterson, W.P. Thurston, *Word Processing in Groups*, Jones and Bartlett, Boston, 1992.

[EHR 91]   D.B.A. Epstein, D.F. Holt and S.E. Rees, *The use of Knuth-Bendix methods to solve the word problem in automatic groups*, J. Symbolic Computation **12** (1991), pp. 397–414.

[Gil 79]   R.H. Gilman, *Presentations of groups and monoids*, J. Alg. **57** (1979), pp. 544–554.

[HaK 84]   G. Havas, L.G. Kovács, *Distinguishing eleven crossing knots*, in: Computational Group Theory (M. Atkinson, ed.), Academic Press, London, 1984, pp. 367–373.

[HaN 80]   G. Havas, M.F. Newman, *Applications of computers to questions like those of Burnside,* in: Burnside Groups (J.L. Mennicke, ed.), Lecture Notes in Mathematics 806, Springer, 1980, pp. 211–230.

[Hav 74]   G. Havas, *A Reidemeister-Schreier program,* in: Proceedings of the Second International Conference on the Theory of Groups, Canberra 1973 (M.F. Newman, ed.), Lecture Notes in Mathematics 372, Springer, Berlin, 1974, pp. 347–356.

[HoR 92a]   D. F. Holt, S. Rees, *Testing for isomorphism between finitely presented groups,* to appear in Proceedings of Conference Groups and Combinatorics, Durham 1990.

[HoR 92b]   D. F. Holt, S. Rees, *A graphics sytem for displaying finite quotients of finitely presented groups,* submitted to Proceedings of DIMACS conference on Computational Group Theory, Rutgers 1991.

[Kel 90]   A.J. Kelly, *Groups from link diagrams, Ph.D. Thesis,* University of Warwick, 1990.

[LMR 92]   K. Lux, J. Müller and M. Ringe, *Peakword condensation and submodule lattices, an application of the Meat-Axe,* to appear.

[Par 84]   R. Parker, *The computer calculation of modular characters. (The Meat-Axe),* in: Computational Group Theory (M. Atkinson, ed.), Academic Press, London, 1984, pp. 267–74.

Proving Certain Groups Infinite

James Howie, Richard M. Thomas

Department of Mathematics, Heriot-Watt University, Edinburgh, EH14 4AS.
Department of Computing Studies, University of Leicester, Leicester, LE1 7RH.

The group $(2, 3, r; s)$ is defined by the presentation

$$\langle a, b \mid a^2 = b^3 = (ab)^r = [a, b]^s = 1 \rangle.$$

The question that interests us is that of deciding which of the groups $(2, 3, r; s)$ are finite. The aim of this paper is to briefly survey what is known about these groups and then to highlight the one remaining problem.

To begin with, the groups $(2, 3, 3; s)$, $(2, 3, 4; s)$ and $(2, 3, 5; s)$ are homomorphic images of the triangle groups $(2, 3, 3)$, $(2, 3, 4)$ and $(2, 3, 5)$ respectively, and hence are all finite, and $(2, 3, 6; s)$ is a non-cyclic finite group of order $6s^2$. The group $(2, 3, r; 2)$ is isomorphic to A_4 if $r \equiv 3 \pmod 6$, $A_4 \times C_2$ if $r \equiv 0 \pmod 6$, and is trivial otherwise, and the group $(2, 3, r; 3)$ has order $\frac{3}{2}r^2$ if r is even, and is trivial otherwise [Sin36]. So we will assume that $r \geq 7$ and $s \geq 4$.

It was pointed out in [Cox39] that the group $(3, 3 \mid u, s)$ with presentation

$$\langle c, d \mid c^3 = d^3 = (cd)^u = (c^{-1}d)^s = 1 \rangle$$

has index 2 in $(2, 3, 2u; s)$, and is thus infinite when $(2, 3, 2u; s)$ is infinite. In turn, $(3, 3 \mid u, 3v)$ contains $(u, u, u; v)$ as a subgroup of index 3, so that this group is also infinite when $(2, 3, 2u; 3v)$ is infinite. Also, $(2, 3, r; s)$ has index 2 in the group $G^{3,r,2s}$ with presentation

$$\langle r_1, r_2, r_3 \mid r_1^2 = r_2^2 = r_3^2 = (r_1 r_2)^2 = (r_2 r_3)^3 = (r_3 r_1)^r = (r_1 r_2 r_3)^{2s} = 1 \rangle,$$

so that $G^{3,r,2s}$ is infinite when $(2, 3, r; s)$ is infinite. In particular, since $G^{3,m,n}$ is isomorphic to $G^{3,n,m}$, $(2, 3, 2k; m)$ is infinite if and only if $(2, 3, 2m; k)$ is infinite.

Now $(2, 3, 7; 4)$ is isomorphic to $PSL(2, 7)$, $(2, 3, 7; 5)$ is trivial, while $(2, 3, 7; 6)$ is isomorphic to $PSL(2, 13)$ [Bra28], as is $(2, 3, 7; 7)$ [Sin35]. The group

$(2,3,8;4)$ is isomorphic to $PGL(2,7)$ [Cox39], while $(2,3,8;5)$ and $(2,3,10;4)$ have a central subgroup of order 3 with factor group A_6 extended by an outer automorphism of S_6 [Cox39], and $(2,3,8;6)$ is infinite [Bra31]. The group $(2,3,9;4)$ is isomorphic to A_4 [Cox39] and $(2,3,9;5)$ was thought to be cyclic of order 3 (though we will see in a moment that this is incorrect). The group $(2,3,11;4)$ is isomorphic to $PSL(2,23)$ [Cox39].

It was shown in [Cox57] that, if m and n are both even, then

$$G^{3,m,n} \text{ finite } \iff \cos\left(\frac{4\pi}{m}\right) + \cos\left(\frac{4\pi}{n}\right) < \frac{1}{2} \qquad (*)$$

and so, for even r, $(2,3,r;s)$ is infinite if and only if $\cos\left(\frac{4\pi}{r}\right) + \cos\left(\frac{2\pi}{s}\right) < \frac{1}{2}$. Leech and Mennicke [LeM61] showed that $(2,3,7;8)$ is an extension of an elementary abelian group of order 2^6 by $PSL(2,7)$; see also [Cox62]. $(2,3,9;5)$ was shown to be a direct product of C_3 with $PSL(2,19)$ [Sin69] (as opposed to being a cyclic group). On the other hand, Sims [Sim64] and Leech [Lee66] showed that $(2,3,7;9)$ is infinite; see also [Hig], where it is also shown that $(2,3,7;10)$ is infinite, and some further results may be found there and in [GrM91]. The following result was proved in [HoT90], [HoT]:

Theorem A. *Let r and s be integers satisfying one of the following conditions:*

(i) $s = 4$ and $r \geq 15$;   (ii) $s = 5$ and $r \geq 10$;   (iii) $s \geq 6$ and $r \geq 9$;
(iv) $s \geq 8$ and $r \geq 8$;   (v) $s \geq 12$ and $r \geq 7$.
Then the group $(2,3,r;s)$ is infinite.

An independent proof for $r = 7$ and $s \geq 12$ may be found in [HoP], and the group $(2,3,7;11)$ has now been shown to be infinite also by Martin Edjvet [Edj]. So Theorem A, together with the results mentioned above, yields:

Theorem B. *With the possible exception of $(r,s) = (13,4)$, the group $(2,3,r;s)$ is infinite if and only if r and s satisfy one of the following conditions:*

(i) $r = 7$, $s \geq 9$;   (ii) $r = 8$ or 9, $s \geq 6$;   (iii) $r = 10$ or 11, $s \geq 5$;
(iv) $r \geq 12$, $s \geq 4$.

Theorem B shows that, if n is even, then $(*)$ is necessary and sufficient for arbitrary m with the possible exception of $m = 13$, $n = 8$. While we do not know whether or not $(2,3,13;4)$ is infinite, it was shown in [Sin37] that the homomorphic image of $(2,3,13;4)$ defined by the presentation

$$\langle a,b \mid a^2 = b^3 = (ab)^{13} = [a,b]^4 = w(a,b) = 1 \rangle$$

is isomorphic to $PSL(3,3)$, where, if d and e are defined to be ab^{-1} and $babab$ (so that $d^{13} = e^4 = 1$), then $w(a,b)$ is the word $ed^7ed^7e^2d^{-7}e^2d^7$. The homomorphism from $(2,3,13;4)$ onto $PSL(3,3)$ gives a permutation representation of $(2,3,13;4)$ of degree 13, which may be realized as follows:

$$a \rightarrow (2,3)(4,13)(5,11)(7,10),$$
$$b \rightarrow (1,2,4)(5,12,13)(6,7,11)(8,9,10),$$
$$ab \rightarrow (1,2,3,4,5,6,7,8,9,10,11,12,13),$$
$$[a,b] \rightarrow (1,5,8,11)(2,4,13,3)(6,12)(7,10).$$

We then have that $ab^{-1}abab \rightarrow (2,5,9,10,8,12,7,11)(1,6,13,4)$, so that the relation $(ab^{-1}abab)^8 = 1$ also holds in $PSL(3,3)$; we shall see in a moment that this extra relation, together with those defining $(2,3,13;4)$, are enough to define $PSL(3,3)$.

Clearly any finite homomorphic image of $(2,3,13;4)$ must be perfect. Holt has checked, using a computer, that the only finite simple groups of order less than 10^6 that are homomorphic images of $(2,3,13;4)$ are $PSL(3,3)$ and $PSL(2,25)$, and that $(2,3,13;4)$ also has as a homomorphic image a group E which is an extension of an elementary abelian group of order 2^{12} by $PSL(3,3)$, so that $(2,3,13;4)$ has homomorphic image $E \times PSL(2,25)$. Further to this, George Havas and Edmund Robertson have determined, again using a computer, that the presentations

$$\langle a,b \mid a^2 = b^3 = (ab)^{13} = [a,b]^4 = (ab^{-1}abab)^8 = 1 \rangle,$$
$$\langle a,b \mid a^2 = b^3 = (ab)^{13} = [a,b]^4 = (ab^{-1}abab)^{16} = 1 \rangle,$$
$$\langle a,b \mid a^2 = b^3 = (ab)^{13} = [a,b]^4 = (ab^{-1}abab)^{13} = 1 \rangle.$$

determine $PSL(3,3)$ (as we mentioned above), E and $PSL(2,25)$ respectively. Here we will verify Holt's result, and extend it slightly, by proving:

Theorem C. *There is no non-trivial homomorphism of $(2,3,13;4)$ into $GL(2,F)$ for any field F. If θ is a non-trivial homomorphism of $(2,3,13;4)$ into $PGL(2,F)$ for some field F, then F has characteristic 5 and the image of θ is isomorphic to $PSL(2,25)$.*

With regards to the simple groups of order less that 10^6, note that, since $(2,3,13;2)$ is trivial, the images of a, b, ab and $[a,b]$ in any non-trivial homomorphic image of $(2,3,13;4)$ must have orders 2, 3, 13 and 4 respectively. Since the image of $ab^{-1}ab$ has order 4, the images of a and $b^{-1}ab$ must generate a dihedral group of order 8. So we are only interested in finite simple groups of order less than 10^6 whose orders are divisible by $8.3.13 = 312$; this leaves the following possibilities:

$$PSL(3,3); \ PSL(2,25); \ PSU(3,4); \ PSL(2,79); \ PSL(2,64); \ PSL(2,103).$$

Of these, we can immediately rule out $PSU(3,4)$, as a Sylow 2-subgroup only contains three involutions, and hence cannot contain a dihedral subgroup of order 8. Since we already know that $PSL(3,3)$ is a homomorphic image of $(2,3,13;4)$, the rest will follow from Theorem C.

We start by recalling some basic facts about $GL(2, F)$ that will be useful. Consider $GL(2, F)$ acting on a vector space V of dimension 2 over F. If R is any element in $GL(2, F)$ which is not a scalar multiple of the identity I, u is any non-zero vector which is not an eigenvector of R and $v := Ru$, then $B := \{u, v\}$ is a basis for V and R acts as $\begin{pmatrix} 0 & 1 \\ s & t \end{pmatrix}$ with respect to B. If R has determinant 1, then $s = -1$, so that R is conjugate to $\begin{pmatrix} 0 & 1 \\ -1 & t \end{pmatrix}$ in $GL(2, F)$. Since conjugate matrices have the same trace, the matrices $\begin{pmatrix} 0 & 1 \\ -1 & t \end{pmatrix}$ form a complete set of representatives for the distinct conjugacy classes of non-scalar matrices of determinant 1 in $GL(2, F)$; a consequence is that $-I$ is the only element of order 2 in $SL(2, F)$. Since I and $-I$ are the only scalar matrices in $SL(2, F)$, the order of any matrix of trace other than ± 2 is uniquely determined by its trace; if we are working in $PSL(2, F)$, this still holds, subject to the fact that the trace of a matrix is now only defined up to multiples of ± 1, since M and $-M$ are equal in $PSL(2, F)$.

With regards to the proof of Theorem C, it is clear that, if θ is any homomorphism of $(2, 3, 13; 4)$ into $GL(2, F)$ (or $PGL(2, F)$), then the image of θ is perfect, and so all matrices in the image of θ have determinant 1 (or ± 1 in the case of $PGL(2, F)$); so we have a homomorphism from $(2, 3, 13; 4)$ into $SL(2, F)$ or $PSL(2, F)$. We cannot have a non-trivial homomorphism into $SL(2, F)$, since $-I$ is the only element of order 2 in $SL(2, F)$; therefore let us consider elements A and B of $PSL(2, F)$ such that $A^2 = B^3 = (AB)^{13} = [A, B]^4 = 1$. Considering A as an element of $SL(2, F)$, we have that $A^2 = -I$, and, replacing A by a conjugate if necessary, we may take A to be $\begin{pmatrix} 0 & -1 \\ 1 & 0 \end{pmatrix}$. Since any matrix of order three in $PSL(2, F)$ has trace ± 1, B must be of the form $\begin{pmatrix} \alpha & \beta \\ \gamma & 1 - \alpha \end{pmatrix}$ (replacing B by $-B$ if necessary), where $\alpha - \alpha^2 - \beta\gamma = 1$. Let $\tau := \beta - \gamma$; then

$$AB = \begin{pmatrix} -\gamma & \alpha - 1 \\ \alpha & \beta \end{pmatrix} \quad \text{and} \quad A^{-1}B^{-1}AB = \begin{pmatrix} \alpha^2 + \gamma^2 & \alpha\tau + \gamma \\ \alpha\tau + \gamma & (\alpha - 1)^2 + \beta^2 \end{pmatrix},$$

so that $C := AB$ has trace τ and $D := A^{-1}B^{-1}AB$ has trace $2\alpha^2 - 2\alpha + 1 + \beta^2 + \gamma^2 = \tau^2 - 1$.

We may readily check that, if ω is the trace of an element of order 4 in $PSL(2, F)$, then $\omega^2 = 2$; so, if $D^4 = 1$, we must have that $\tau^2 = \omega + 1$. Now, in $SL(2, F)$, $\operatorname{Tr}(UV) = \operatorname{Tr}(U)\operatorname{Tr}(V) - \operatorname{Tr}(U^{-1}V)$, and so, putting $U = C$ and $V = C^{m-1}$, we get that $\operatorname{Tr}(C^m) = \tau\operatorname{Tr}(C^{m-1}) - \operatorname{Tr}(C^{m-2})$. This gives that C^2 has trace $\tau^2 - 2 = \omega - 1$, C^3 has trace $\tau(\omega - 2)$, C^4 has trace $1 - 2\omega$, etc., and we may continue this to deduce that C^{13} has trace $\tau(47 - 34\omega)$ in $SL(2, F)$. So, if C has order 13 in $PSL(2, F)$, we must have that $\tau(47 - 34\omega) = \pm2$, so that $\tau^2(47 - 34\omega)^2 = 4$, which gives that $1325\omega = 1875$; so F has characteristic $p > 0$. If $p \neq 5$, we have $53\omega = 75$, and squaring both sides yields that $5618 = 5625$, so that $p = 7$.

If $p = 5$, then $\tau = \pm(\omega - 2)$, and we may check that $C^{13} = -I$. On the other hand, if $p = 7$, then $\omega = 3$ or -3, and then C^{13} has trace $\tau(5 + \omega) = \tau$ or 2τ respectively. So $\tau = \pm2$ or $\tau = \pm1$, so that $\tau^2 = 4$ or 1, and, since $\tau^2 = \omega + 1$, we must have $\tau = \pm2$, $\omega = 3$. But now AB is conjugate to $\begin{pmatrix} 0 & 1 \\ -1 & \pm2 \end{pmatrix}$, an element of $PSL(2, 7)$, contradicting the fact that $PSL(2, 7)$ has no element of order 13; thus $p = 5$. Since D has trace ω, and hence order 4, we now have a homomorphic image of $(2, 3, 13; 4)$.

With A and B as above (and F of characteristic 5), we have that $A^2 = B^3 = -I$ in $SL(2, F)$. Since

$$\mathrm{Tr}\,(AB^{-1}C^2) = \mathrm{Tr}\,(AB^{-1})\mathrm{Tr}\,(C^2) - \mathrm{Tr}\,(BA^{-1}C^2),$$

and

$$\mathrm{Tr}\,(C^2) = \omega - 1,$$
$$\mathrm{Tr}\,(AB^{-1}) = \mathrm{Tr}\,(BA^{-1}) = -\mathrm{Tr}\,(BA) = -\mathrm{Tr}\,(C) = -\tau = \mp(\omega - 2),$$
$$\mathrm{Tr}\,(BA^{-1}C^2) = \mathrm{Tr}\,(B^2AB) = \mathrm{Tr}\,(AB^3) = -\mathrm{Tr}\,(A) = 0,$$

we have that

$$\mathrm{Tr}\,(AB^{-1}C^2) = \mp(\omega - 2)(\omega - 1) = \pm(1 - 2\omega) = \pm\mathrm{Tr}\,(C^4).$$

Since C^4 has order 13 in $PSL(2, F)$, we have that $(AB^{-1}C^2)^{13} = \pm I$, and we have the relations

$$A^2 = B^3 = (AB)^{13} = [A, B]^4 = (AB^{-1}ABAB)^{13} = 1,$$

which is a complete set of defining relations for $PSL(2, 25)$ by the above. So $< A, B >$ is isomorphic to $PSL(2, 25)$, and we have completed the proof of Theorem C.   □

Acknowledgement. The second author would like to thank Hilary Craig for all her help and encouragement.

References

[Bra28] H. R. Brahana, *Certain perfect groups generated by two operators of orders two and three*, Amer. J. Math. **50** (1928), pp. 345–356.

[Bra31] H. R. Brahana, *On the groups generated by two operators of orders two and three whose product is of order 8*, Amer. J. Math. **53** (1931), pp. 891–901.

[Cox39] H. S. M. Coxeter, *The abstract groups $G^{m,n,p}$*, Trans. Amer. Math. Soc. **45** (1939), pp. 73–150.

[Cox57] H. S. M. Coxeter, *Groups generated by unitary reflections of period two*, Canadian J. Math. **9** (1957), pp. 243–272.

[Cox62] H. S. M. Coxeter, *The abstract group $G^{3,7,16}$*, Proc. Edinburgh Math. Soc. **13** (1962), pp. 47–61 and 189.

[Edj] M. Edjvet, *An example of an infinite group*, preprint.

[GrM91] L. C. Grove and J. M. McShane, *On Coxeter's groups $G^{p,q,r}$*, in: Groups St. Andrews 1989, vol. 1 (C. M. Campbell and E. F. Robertson, eds.), L.M.S. Lecture Note Series 159, Cambridge University Press, Cambridge, 1991 pp. 211–213.

[Hig] G. Higman, *The groups $G^{3,7,n}$*, unpublished notes, Mathematical Institute, University of Oxford.

[HoP] D. F. Holt, W. Plesken, *A cohomological criterion for a finitely presented group to be infinite*, preprint.

[HoT90] J. Howie and R. M. Thomas, *On the asphericity of presentations for the groups $(2,3,p;q)$ and a conjecture of Coxeter*, Technical Report 36, Department of Computing Studies, University of Leicester, May 1990.

[HoT] J. Howie and R. M. Thomas, *The groups $(2,3,p;q)$; asphericity and a conjecture of Coxeter*, to appear in J. Algebra.

[Lee66] J. Leech, *Note on the abstract group $(2,3,7;9)$*, Proc. Cambridge Phil. Soc. **62** (1966), pp. 7–10.

[LeM61] J. Leech and J. Mennicke, *Note on a conjecture of Coxeter*, Proc. Glasgow Math. Assoc. **5** (1961), pp. 25–29.

[Sim64] C. C. Sims, *On the group $(2,3,7;9)$*, Notices Amer. Math. Soc. **11** (1964), pp. 687–688.

[Sin35] A. Sinkov, *A set of defining relations for the simple group of order 1092*, Bull. Amer. Math. Soc. **41** (1935), pp. 237–240.

[Sin36] A. Sinkov, *The group defined by the relations $S^l = T^m = (S^{-1}T^{-1}ST)^p = 1$*, Duke Math. J. **2** (1936), pp. 74–83.

[Sin37] A. Sinkov, *Necessary and sufficient conditions for generating certain simple groups by two operators of periods two and three*, Amer. J. Math. **59** (1937), pp. 67–77.

[Sin69] A. Sinkov, *The number of abstract definitions of $LF(2,p)$ as a quotient group of $(2,3,n)$*, J. Algebra **12** (1969), pp. 525–532.

Some Applications of Small Cancellation Theory to One-Relator Groups and One-Relator Products

Arye Juhász

Department of Mathematics, Technion-Israel Institute of Technology, 32000 Haifa, Israel.

Introduction

It was believed for some time that most of the one-relator groups would be small cancellation groups. In this talk I would like to confirm this by showing that, with the exception of a small class of one-relator groups, they are small cancellation groups, and the groups in the exceptional class are small cancellation in a generalized sense, with one exception, where they are only small cancellation groups relative to a subgroup. These results are sufficient in order to regard one-relator groups as if they were small cancellation groups. In particular, we get a solution for the conjugacy problem for all one-relator groups. More precisely, it is shown that Schupp's solution to groups satisfying one of the geometrical small cancellation conditions [5] applies to one-relator groups with one exceptional class, where it is possible to reduce it to a combinatorial problem which cannot be solved by small cancellation theory but can be solved by a very simple combinatorial consideration [4].

1. The non-small cancellation groups

Let $P = \langle X \mid R \rangle$ be a one-relator presentation, R cyclically reduced. If P satisfies the condition $C(6)$ then it is, by definition, small cancellation. So we shall assume that

$$P \text{ does not satisfy the condition } C(6).$$

This means that there is a connected, simply connected, reduced in the sense of Lyndon, Schupp [5] diagram M over $\mathcal{R}$ (the symmetrical closure of R) which contains an inner region with less than six neighbours (i.e. a polygon with less than six edges). For every presentation P denote by $m(P)$ the minimal number of neighbours which an inner region may have:

$$m(P) = \min_{M \in \mathcal{M}} |\{d(D) \mid D \in \operatorname{Inreg}(M)\}|.$$

Here $\mathcal{M}$ stands for the set of all the connected and simply connected diagrams over $\mathcal{R}$, D is a region with $d(D)$ edges and $\operatorname{Inreg}(M)$ stands for the set of all the inner regions of M.

We have to deal with the following four cases:

1. $m(P) = 2$,   3. $m(P) = 4$,
2. $m(P) = 3$,   4. $m(P) = 5$.

Clearly the groups occurring in cases 2, 3 and 4 still may be small cancellation groups if the vertices of the corresponding regions have high enough valency. However, at the moment from our point of view the important thing is that all the non-small cancellation groups are covered by these cases.

Let us consider them more closely. If an inner region has m neighbours, then one of the cyclic conjugates of R, say R itself, can be decomposed as the product of m subwords $P_1, \ldots, P_m$ in such a way that either the words P_i or their inverses are subwords of the relators which are represented by the neighbouring regions. Since we have only one relator, the boundary label of every region is a cyclic conjugate of R or of R^{-1}. Consequently $R = P_1 \ldots P_m$ and each P_i satisfies one of the following equations:

$$(*) \begin{cases} U_i P_i' V_i = R, & P_i' = P_i^{\varepsilon_i}, & \varepsilon_i = \pm 1, \\ U_i \neq P_1 \ldots P_{i-1}, & (P_0 = 1), \\ P_i' = H_i T_i, & T_i U_i H_i = R, & P_i' = P_i, & H_i \neq 1 \neq T_i, & i = 1, \ldots, m. \end{cases}$$

The solutions to these equations give all the words which are suspected of giving rise to a non small-cancellation group. Thus, if there would exist something like "algebraic geometry over free groups" then the set of small cancellation groups would contain an "open subset" in the corresponding "Zariski Topology". This would give some confirmation to the feeling that most of the one-relator groups are small cancellation groups.

2. The extended small cancellation theory

We now develop the necessary extension of small cancellation theory which enables us to deal with these groups. All solutions of $(*)$ which classify the non-small cancellation cases, can be expressed as words in subwords of R on X. We explain the main ideas through the following example. Assume $m = 2$, $P_1 = A$, $P_2 = B$, $|A| \geq |B|$; then one of the solutions to the equations is $A = (ST)^\alpha S$, $\alpha \geq 1$, $B = TS$, $R = (ST)^{\alpha+1}S$, where S and T are arbitrary words on X such that $(ST)^{\alpha+1}S$ is cyclically reduced. To simplify this discussion we shall assume $\alpha = 1$; thus $A = STS$, $B = TS$, $U_1 = ST$, $V_1 = 1$ $U_2 = S$, $V_2 = TS$ and $R = STSTS$. We shall call these three occurrences of S and these two occurrences of T their *standard occurrences* in R. Let M be a fixed simply connected $\mathcal{R}$-diagram. In order to avoid inner regions with two neighbours, we construct a new presentation out of the given one by adding further relations in the following way: Consider R as a word in the subgroup $\langle S, T \rangle = H$. The relations we add are all the words in the normal closure of $\langle R \rangle$ in H and we regard these relations as the *relators* in our new presentation. Geometrically this is carried out as follows: say that two regions are H-*adjacent* if the label of their common edge is a product of standard occurrences of S and T (or their inverses), regarding a single occurrence of $S^{\pm 1}$ or $T^{\pm 1}$ as a product. Call the regions D_0, D_t H-*linked* if there is a sequence of regions $D_1, \ldots, D_{t-1}$ such that D_i and D_{i+1} are H-adjacent for $i = 0, \ldots, t - 1$. H-linkedness is an equivalence relation which we shall denote by $\underset{\tilde{H}}{\sim}$. Next, for each $D \in \operatorname{Reg}(M)$ define $\Sigma_H(D)$ to be the $\underset{\tilde{H}}{\sim}$-class of D and denote by $\Delta_H(D) = \operatorname{Int}(\bigcup \bar{E})$ where $E \in \Sigma_H(D)$ and $\operatorname{Int}(X)$ is the interior of X.

Assume that the following hypothesis holds:

$(\mathcal{H})$ For every $D \in \operatorname{Reg}(M)$, $\Delta_H(D)$ is simply connected.

(It is proved by induction at the end of the construction that in fact $(\mathcal{H})$ holds (see [6])).

Under $(\mathcal{H})$ we may consider $\Delta_H(D)$ as regions for the presentation $\langle X \mid R^H \rangle$ and we get a diagram over $\langle R \rangle^H$ with boundary ∂M. We shall call this diagram the *derived diagram of M with respect to H*, and denote it by M_H. This whole construction is a very much simplified version of a construction due to E. Rips [6].

We shall call $\Delta_H(D)$ a *derived region of M with respect to H* and denote the set of all the derived regions by $\operatorname{Reg}_H(M)$.

Let us consider more closely the derived diagram M_H. If $\Delta_1, \Delta_2 \in \operatorname{Reg}_H(M)$, then $\partial \Delta_1 \cap \partial \Delta_2$ cannot be labeled by standard occurrences of S or T. Hence, if all the occurrences of $S^{\pm 1}$ and $^{\pm 1} T$ in the cyclic conjugates of R are standard, then $|\partial \Delta_1 \cap \partial \Delta_2| < |ST|$. If, in addition, we assume that

$(\mathcal{A})$ ST and TS are the product of at least four pieces, relative to the symmetric closure of ST,

then Δ has at least six neighbours. Hence in this case M_H satisfies the condition $C(6)$ (in spite of the fact that the original presentation does not satisfy even $C(3)$). In this situation we say that P is a *small cancellation presentation with respect to H*. This very much resembles Gromov's Relative Hyperbolicity (§8.6, p. 256 in [2], Example (b) in particular.) In order to make this construction applicable we still have to consider the case when the whole diagram M_H consists of a single region Δ. In this case, of course, relative small cancellation does not help. But in our example, as a simply connected diagram over R, Δ turns out to be a diagram without inner regions; therefore by definition it is small-cancellation. This is not a surprise, for $\widetilde{H}$, the image of H in the group defined by P, turns out to be isomorphic to $\langle x, y \mid xyxyx \rangle$, which is isomorphic to $\mathbb{Z}$, and $\mathbb{Z}$ is hyperbolic. Having these two results, namely, that P is small cancellation relative to H and H is small cancellation, we may regard M_H as a small cancellation diagram. In particular we can solve the conjugacy problem by P. Schupp's method [5].

3. Solution of the conjugacy problem

First we recall Schupp's method very briefly in a way suited to our purpose. Let u and v be cyclically reduced words. Define $c^1(u) = \{u^h \mid |h| < |R|, \; |u^h| \le |u|\}$. For $k \ge 2$ define inductively

$$c^k(u) = \bigcup_{x \in c^{k-1}(u)} c^1(x).$$

Let $c(u) = \bigcup_{k=1}^{\infty} c^k(u)$ and define $c(v)$ similarly. Clearly $c(u)$ and $c(v)$ are finite sets, and if $c^{k+1}(u) = c^k(u)$ then $c^{k+t}(u) = c^k(u)$, for every $t \ge 1$. Consequently the sets $c(u)$ and $c(v)$ can be found recursively. If P is a small cancellatin presentation, then u is conjugate to v if and only if $c(u) \cap c(v) \ne \emptyset$. Now we turn to our example. Let A be an annular diagram over R with boundaries labeled by u and v and let A' be its derived diagram as constructed above. It satisfies the condition $C(6)$ in terms of the derived regions. Hence Schupp's method will apply if the following condition are satisfied:

(a) There is a closed simple path η in A', homotopic to each of the boundary components of $\partial A'$ such that the regions Δ between η and $\partial A'$ are arranged in layers having a non-decreasing number of regions toward the boundaries.

(b) For every such annular layer L and region Δ in L, $|\Delta L \cap \partial \Delta|$ is bounded by $k|R|$ for a constant k depending only on R.

(c) There exist adjacent regions Δ_1 and Δ_2 in L with $|\partial \Delta_2 \cap \partial \Delta_2| < |R|$.

(d) H has solvable conjugacy problem.

If we choose u and v to be Dehn-reduced words over R (which we certainly can), then it follows from the theory of $W(6)$-diagrams that (a) and (b) are satisfied [2]. Moreover, it follows from assumption $\mathcal{H}$ that (c) holds. Clearly H has a solvable conjugacy problem. Consequently, the following modification of Schupp's procedure applies to our group if we require u and v to be Dehn-reduced.

Check whether $c(u) \cap c(v) \ne \emptyset$. If it is not empty, then certainly u is conjugate to v. If it is empty then check whether $C(u)$ contains an element u of H and

$c(v)$ contains an element v of H such that u and v are conjugate in H. If there are such elements then u and v are conjugate. If no such elements exist, then they are not conjugate.

One advantage of this solution is that if P is small cancellation with respect to H and H has a solvable conjugacy problem, then P has a solvable conjugacy problem. In particular, if H is very bad from the small cancellation theoretical point of view (i.e. it has a non-polynomial isoperimetric inequality) but still has a solvable conjugacy problem due to some combinatorial or group theoretical reasons, then P has a solvable conjugacy problem. In the family of one relator groups, such a phenomenon occurs in the class of Baumslag-Solitar type groups $K = \langle X \mid AB^\alpha A^{-1} B^{-\beta}, \ \alpha \neq \beta \rangle$. Thus, in spite of the fact that K has a very bad isoperimetric inequality (see [1]), K still has a solvable conjugacy problem since K turns out to be small cancellation with respect to $\langle A, B \rangle$ and $\langle x, y \mid xy^\alpha x^{-1} y^{-\beta} \rangle$ has a solvable conjugacy problem due to a completely different combinatorial reason (see [4]). In other words, in these groups the "bad" regions occur in "aggregates" (the derived regions) in such a way that the common boundary of any two such aggregates is a very short (or empty) path.

Let us come back to our assumption $(\mathcal{A})$. If it is not satisfied, then we get new equations of a type similar to $(*)$ and their solution gives finer information on the structure of S and T. This implies that in the new derived diagram with respect to the subgroup generated by these finer pieces (i.e. shorter subwords), the common boundary of two-derived regions will be shorter. Thus iterating this procedure several times (≤ 3) we finally get a derived diagram for which the condition $C(6)$ holds. A more delicate analysis will lead to a complete classification of all words which give rise to one relator groups which are semi-hyperbolic but not hyperbolic. More precisely, this leads to a computer program which, for a given cyclically reduced word, will give one of the three outputs: hyperbolic, semi-hyperbolic, else. This work is in progress.

4. One relator products

Let us turn to one-relator products. This is also a work in progress and I shall be very brief. The method described above applies to one-relator products. The difficulty is that in the one-relator product, the equations obtained have many more solutions than in the one-relator case. This is due to the fact that a word over a free product may overlap with its inverse. Hopefully, this method will give a finite list of words for which the one-relator products are not small cancellation and, by applying the theory of derived diagrams, this list will reduce to a small class of groups defined by generators and relators in terms of the generators in an explicit way (i.e. not just by general words).

Acknowledgements. I would like to express my deep gratitude to Prof. E. Rips for numerous discussions and for his useful suggestions relating to the material in this paper. I am grateful to the referee for his valuable remarks.

References

[1] S.M. Gersten, *Dehn functions and L_1-norms of finite presentations,* in: Proceedings of the Workshop on Algorithmic Problems (G. Baumslag and C.F. Miller III, eds.), MSRI Series 23, Springer, 1991.

[2] M. Gromov, *Hyperbolic Groups,* in: Essays in Group Theory (S.M. Gersten , ed.), MSRI Series 8, Springer, 1987.

[3] A. Juhász, *Small cancellation theory with a unified small cancellation condition,* Journ. London Math. Soc. 40 (1989), pp. 57–80.

[4] A. Juhász, *Solution of the conjugacy problem in one-relator groups,* in: Proceedings of the Workshop on Algorithmic Problems (G. Baumslag and C.F. Miller III, eds.), MSRI Series 23, Springer, 1991.

[5] R.C. Lyndon and P.E. Schupp, *Combinatorial Group Theory,* Springer, 1977.

[6] E. Rips, *Generalized small cancellation theory and applications I, The word problem,* Israel Journal of Math. 41 (1982), pp. 1–146.

A Group Theoretic Proof of the Torus Theorem

Peter H. Kropholler

School of Mathematical Sciences, Queen Mary and Westfield College, Mile End Road, London E1 4NS.

1. Introduction

The Torus Theorem was discovered by Waldhausen and subsequently several authors, including Jaco, Shalen, Johannson and Scott gave proofs of various forms of the Theorem. In its most general form, it identifies a characteristic submanifold within a compact 3-manifold, and provides a canonical decomposition of the 3-manifold into pieces which, according to Thurston's Geometrization Conjecture, admit a geometric structure based on one of eight 3-dimensional geometries. Although Thurston's Conjecture is not at present completely proved, much of it has been proved and it has therefore been clear for several years that the Torus Theorem and its generalizations are of fundamental importance.

The early approaches to proving the Torus Theorem involved quite intricate geometric or topological arguments. Of these, Scott's account [22] and [24] is amongst the most easily digestible and indicates that much of the Theorem depends solely on properties of the fundamental group.

Nowadays any approach to studying a compact 3-manifold M begins by assuming that M is endowed with either a PL or with a smooth structure. It is known that every topological 3-manifold admits both a unique PL and a unique smooth structure, and we shall make no further comment on this part of the theory. A modern approach to the Torus Theorem can begin in one of two ways. Working with the smooth structure, Casson now has a very elegant geometric proof which uses least area surface methods developed by Scott. On the other hand, as we shall demonstrate in this article, one can also work almost entirely with the fundamental group. From this second point of view it is natural to think in terms of the PL structure.

For the sake of simplicity we shall address the following special case of the Torus Theorem:

The Torus Theorem. *Let M be a closed orientable aspherical 3-manifold. Suppose that the fundamental group $\pi_1(M)$ has a rank 2 free abelian subgroup. Then one of the following holds:*

(i) M admits a 2-sided embedded incompressible torus; or

(ii) $\pi_1(M)$ has an infinite cyclic normal subgroup.

The link with group theory depends on the general fact that if M is a closed orientable aspherical n-manifold then $G = \pi_1(M)$ satisfies Poincaré duality: that is, for any G-module V and any integer i there are isomorphisms

$$H^i(G, V) \cong H_{n-i}(G, V). \qquad (1.1)$$

A group G which satisfies (1.1) is called an orientable n-dimensional Poincaré duality group. The Torus Theorem has a formulation for non-orientable 3-manifolds and a general definition can be given for Poincaré duality groups which includes the non-orientable case. The main goal of this paper is to prove the following result, from which it is fairly straightforward to deduce the Torus Theorem as stated above. In the statement we write $\mathfrak{X}$ for the class of groups H of cohomological dimension 2 which have an infinite cyclic subgroup which is commensurable with all of its conjugates.

Theorem 1.2. *Let G be a 3-dimensional Poincaré duality group. If G has a rank 2 free abelian subgroup then one of the following holds:*

(i) G is a non-trivial free product with amalgamation; $G = K *_H L$, where H belongs to $\mathfrak{X}$; or

(ii) G is an HNN-extension, $G = B*_{H,t}$, where H belongs to $\mathfrak{X}$; or

(iii) G has an infinite cyclic normal subgroup.

Layout of the paper

Section 2. *Deduction of the Torus Theorem from Theorem 1.2*

The deduction depends on a simple and classical tranversality argument, together with an understanding of groups in the class $\mathfrak{X}$.

Section 3. *Poincaré duality groups*

Here we include some background material on Poincaré duality groups.

Section 4. *Ends of pairs of groups*

Two end invariants e and $\tilde{e}$ are defined for a pair of groups $H \leqslant G$, and some splitting theorems for groups are proved. The methods ultimately depend on the ideas of Stallings and Dunwoody concerning bipolar structures and group actions on trees which were first used in the study of groups of cohomological dimension one.

Section 5. *The proof of Theorem 1.2*

Once underway, this proof is quite technical. This should not be surprising
because the theorem is very close to the Torus Theorem and one must expect
it to address similar issues, although from a group theoretic point of view.
There are two fundamental reasons why the Torus Theorem is a hard theorem.
One might like to regard the Torus Theorem as the statement that when the
manifold has an incompressible immersed Torus then it has an incompressible
embedded torus: this presents the first reason to expect problems for there
are counterexamples, first identified by Waldhausen, amongst closed Seifert
fibered manifolds. This is why the Torus Theorem has the exclusion clause in
case G has an infinite cyclic normal subgroup. (By the way, it is now known
that closed irreducible 3-manifolds whose fundamental groups have an infinite
cyclic normal subgroup are Seifert fibered. The analagous question about
PD^3-groups is at present completely unsolved.) Secondly, even when one can
find an embedded torus, it certainly need not be possible to find one which
is freely homotopic to the original torus. Geometrically one expects surgery
techniques to help to build the embedded torus by studying the immersed
torus. Group theoretically we do not have a surgery technique and broadly
speaking, what one has to do is as follows. Let H be a rank 2 free abelian
subgroup of a PD^3-group G. If G does not split as an amalgamation or
HNN-extension over any subgroup commensurable with H then one finds
that there is an infinite cyclic subgroup K of H with very large normalizer.
For technical reasons we have to work with the commensurator C of K rather
than its normalizer, but this is similar in spirit. This new subgroup C contains
an abundance of free abelian groups of rank 2 and the aim is to obtain a
splitting of G over one of these, or failing that, at least over some subgroup of
C. Thus in outline, the group theory mirrors exactly what happens inside a
Haken 3-manifold. The original immersed torus (with fundamental group H)
is freely homotopic to an immersion into some component (with fundamental
group C) of the characteristic submanifold. The boundary components of
this component will be tori or Klein bottles which are incompressible and
embedded.

2. Deduction of the Torus Theorem from Theorem 1.2

The principal result we need, which relates 3-manifold theory to group theory,
goes back to Stallings and Waldhausen. It is also an ingredient in Scott's work
[22] and can be stated as follows:

Theorem 2.1. *Let M be a closed aspherical 3-manifold whose fundamental
group G is either a non-trivial free product with amalgamation, $G = K *_H L$,
or an HNN-extension, $G = K *_{H,t}$, over a subgroup H. Then M admits an*

incompressible 2-sided aspherical surface S and the induced map $\pi_1(S) \longrightarrow$
$\pi_1(M)$ carries $\pi_1(S)$ into a subgroup of a conjugate of H.

Proof. The following argument is due to Dunwoody: it is similar in spirit
to the proof of theorem VI.4.4 of [4]. We suppose that there is a given
PL-triangulation of M. Let $\widetilde{M}$ denote the universal cover of M. The given
decomposition of G as amalgamation or HNN-extension is provides an action
of G on a tree T with one orbit of edges and 2 or 1 orbits of vertices, and such
that H is the stabilizer of an edge, e, say. Thus we have two G-spaces $\widetilde{M}$ and
T, and G acts freely on M. Now one can choose a transverse G-map ϕ from
$\widetilde{M}$ to T starting from any G-map between the 0-cells. Let x be the midpoint
of some edge of T. Then $\phi^{-1}(x)$ is a sub-2-manifold of $\widetilde{M}$ and moreover,
the stabilizer of any component of $\phi^{-1}(x)$ is contained in H. Passing to the
quotient M, we can choose a component of the image of $\phi^{-1}(x)$ which yield
a non-trivial decomposition of M. Let X denote this component. Then the
induced map $\pi_1(X) \longrightarrow \pi_1(M)$ carries $\pi_1(X)$ into H, but it might not be
injective; that is, X might not be incompressible in M. The Loop Theorem
shows that if X is not incompressible then there is an essential loop C in
X which is contractible in M, and there is a disc in M spanning this loop.
Now one can do surgery along this disc to reduce the genus of X: if C is
non-separating in X then surgery yields a new X of smaller genus at once,
and if C separates X then surgery divides X into two components, both of
smaller genus, and one can choose one of these so that it still gives a non-
trivial decomposition of M. Since every sphere in M bounds a ball (M being
aspherical) one can adjust the original map ϕ so that $\phi^{-1}(x)$ actually realizes
this new surface of smaller genus and in this way we can reduce to the case
when X is incompressible, so proving the Theorem.

The need for the Loop Theorem in this proof makes it a non-trivial result.
Moreover, as the Theorem is usually stated for irreducible 3-manifolds, rather
than aspherical, one normally also applies the Sphere Theorem to guarantee
that every 2-sphere in M bounds a ball. For these reasons, Theorem 2.1 is the
main part of our proof of the Torus Theorem which appeals to some genuine
3-manifold theory and so it is interesting to note that Stallings work on ends
and almost invariant sets, which we generalize here for our later arguments,
was originally designed to provide a more algebraic proof of the Loop and
Sphere Theorems.

Taking Theorems 1.2 and 2.1 together all that remains in proving the Torus
Theorem as stated above is to check that the incompressible surface obtained
is a torus. One knows that its fundamental group is a subgroup of an $\mathfrak{X}$-group,
and the result follows from the following structure theorem established in [16].

Theorem 2.2. *Let H be a non-trivial finitely generated $\mathfrak{X}$-group. Then H
is the fundamental group of a finite graph of groups in which all the vertex*

and edge groups are infinite cyclic.

It is a consequence of this theorem that the only closed aspherical surface groups which can be subgroups of $\mathfrak{X}$-groups are the Torus group and the Klein bottle group. We can exclude the Klein bottle on grounds of its non-orientability.

3. Poincaré duality groups

An n-dimensional Poincaré duality group, or PD^n-group, is a group of type (FP) which has cohomological dimension n and such that $H^i(G, \mathbb{Z}G) = 0$ for $i < n$ and $H^n(G, \mathbb{Z}G) \cong \mathbb{Z}$. Since $\mathbb{Z}G$ can be regarded as a bimodule, the cohomology group $H^n(G, \mathbb{Z}G)$ inherits an action of G. We say that G is orientable if G acts trivially on $H^n(G, \mathbb{Z}G)$ and non-orientable otherwise. Let us write D for this cohomology group together with its natural action of G: it is called the dualising module of G and one has the following duality between the cohomology and homology of G:

$$H^i(G, V) \cong H_{n-i}(G, V \otimes D), \qquad (3.1)$$

for any G-module V. In fact (3.1) is equivalent to the definition we have given for a Poincaré duality group: for further information we refer the reader to Bieri's book [1].

All known examples of Poincaré duality groups are fundamental groups of closed aspherical manifolds. If M is a closed manifold then one has classical Poincaré duality between the singular homology and cohomology of M, and this duality holds for any coefficient system, trivial or twisted. If M is also aspherical then it is an Eilenberg-Mac Lane space for its fundamental group $G = \pi_1(M)$. Thus G inherits the Poincaré duality from M. This shows that fundamental groups of closed aspherical manifolds are Poincaré duality groups: moreover they are orientable or non-orientable according as the manifold is orientable or non-orientable. It is a deep and difficult question whether the converse holds. It is fairly easy to prove that a PD^1-group is necessarily infinite cyclic and it is a deep theorem of Eckmann, Bieri, Müller and Linnell that PD^2-groups are surface groups. The results described in this article might be regarded as the first tentative steps towards proving that every PD^3-group is a 3-manifold group.

All Poincaré duality groups are torsion-free. The three most important sources of examples are torsion-free polycyclic-by-finite groups, torsion-free cocompact lattices in semisimple Lie groups [2], and torsion-free subgroups of finite index in certain Coxeter groups [3]. We shall have more to say of Davis' examples [3] shortly.

It is easy to show that every subgroup of finite index in a PD^n-group is again a PD^n-group and that one can obtain the dualising module of the subgroup simply by restricting the action of the whole group. Since the dualising module D of a PD^n-group G is additively isomorphic to $\mathbb{Z}$, either G acts trivially on it or G has a subgroup of index 2 which acts trivially. In particular, every non-orientable PD^n-group has a unique orientable subgroup of index 2.

Subgroups of finite index are thus very well behaved. What can be said of infinite index subgroups? There is one general fact, proved by Strebel [27], which plays a fundamental role:

Theorem 3.2. *Every subgroup of infinite index in a PD^n-group has cohomological dimension strictly less than n.*

For example, every subgroup of infinite index in a PD^2-group has cohomological dimension $\leqslant 1$ and so is free by the Stallings-Swan Theorem [26] and [28].

Aside from Theorem 3.2 there are few restrictions known on the possible subgroups of Poincaré duality groups. For example, the situation is much less well understood for PD^3-groups than for 3-manifold groups. It is known that neither a PD^3-group nor a 3-manifold group can contain a direct product of two non-cyclic free groups but in the PD^3-group case this is an isolated result, proved in [12], whereas for 3-manifolds it is a consequence of Scott's compact submanifold theorem [20].

Recently, in a short note [17], Mess has pointed out a reasonably general way of embedding groups as subgroups of Poincaré duality groups, based on Davis' construction [3] with Coxeter groups. We outline briefly what Mess proves:

Theorem 3.3. *If G is the fundamental group of a compact aspherical triangulated n-manifold M then G can be embedded into a PD^n-group.*

Outline of proof. This is proved by starting from a sufficiently fine triangulation of the boundary of M (the barycentric subdivision of the given triangulation is good enough), and then placing a panel structure on the boundary with one panel for each vertex of the triangulation and with two panels intersecting when the corresponding vertices are joined by an edge of the triangulation. The panel structure is dual to the triangulation, and one imagines the panels being mirrors so that if one stands within the manifold one sees many reflections in the different panels yielding a development of the manifold to a much larger open manifold on which a reflection group is acting. More formally, one defines a Coxeter group with one generating reflection for each vertex of the boundary of M and subject to the relations that two vertices commute if and only if they are joined by an edge, or equivalently if and only if the corresponding panels meet. The development of M, which is again a manifold, consists of a family of copies of M, one for each element of the Coxeter group, with two copies being identified together along the appropriate panel when there corresponding group elements differ by a single reflection. Davis applies this to the case when M is contractible and shows that the development of M is also contractible. In consequence any torsion-free subgroup of finite index in the Coxeter group is a Poincaré duality group because it acts freely and cocompactly on the developed space. Davis points out that more complicated examples can be obtained if one only assumes that M is aspherical, and Mess takes this further: if M is aspherical the developed space is still aspherical and its universal cover admits a discrete properly discontinuous and cocompact action of a group which is an extension of a group involving $G = \pi_1(M)$ by the Coxeter group. In this way G can be embedded into a Poincaré duality group.

Corollary 3.4. *If G is the fundamental group of a finite aspherical k-dimensional complex X then G can be embedded into a PD^{2k+1}-group.*

Proof. Every finite k-dimensional complex is homotopy equivalent to a Euclidean neighbourhood retract in $2k + 1$ dimensional space, and so one can replace X with a $2k + 1$-dimensional manifold with the same fundamental group G. Now Theorem 3.3 applies.

The corollary is a fruitful source of examples, and in particular there is one example noted already by Mess in [17] which is worth repeating here. The Baumslag-Solitar group with presentation

$$G = \langle x, y \; ; \; y^{-1}x^2y = x^3 \rangle$$

has a 2-dimensional Eilenberg-Mac Lane space (which is built in the obvious way by adjoining a single 2-cell to a bouquet of 2-circles) and it can therefore be embedded into a PD^5-group. This shows that there are PD^5-groups which do not satisfy Max-c, the maximal condition on centralisers: the chain

$$C_G(x) < C_G(x^2) < C_G(x^4) < C_G(x^8) < \cdots$$

is a strictly increasing chain of centralisers.

In [15], a group theoretic form of the Torus Decomposition Theorem is proved for Poincaré duality groups which satisfy Max-c. It is conceivable that such a theory can be developed for arbitrary Poincaré duality groups, but as Mess' example shows one will have to find methods which are independent of whether or not Max-c holds.

4. Ends of pairs of groups

Let G be a group and let H be a subgroup. There are two end invariants which one can associate to the pair (G, H), denoted by $e(G, H)$ and $\tilde{e}(G, H)$. The first of these has good geometric interpretations and can often be computed by means of the following result:

Lemma 4.1. *If G is finitely generated then $e(G, H)$ is equal to the number of ends of the quotient graph Γ/H, where Γ is the Cayley graph of G with respect to some finite generating set.*

One can give a purely algebraic definition of $e(G, H)$, which makes sense whether or not G is finitely generated. For a set X let $\mathcal{P}X$ denote the power set of X and let $\mathcal{F}X$ denote the set of finite subsets of X. Both $\mathcal{P}X$ and $\mathcal{F}X$ can be regarded as $\mathbb{F}_2$-vector spaces, with addition being the symmetric difference. If X is a G-set then $\mathcal{P}X$ and $\mathcal{F}X$ become $\mathbb{F}_2G$-modules. To define $e(G, H)$, we can take X to be the set of cosets of H in G:

$$H\backslash G := \{Hg; g \in G\}.$$

Now $e(G, H)$ is the dimension over $\mathbb{F}_2$ of the subspace of G-fixed points in $\mathcal{P}(H\backslash G)/\mathcal{F}(H\backslash G)$:

$$e(G, H) := \dim\left(\mathcal{P}(H\backslash G)/\mathcal{F}(H\backslash G)\right)^G. \tag{4.2}$$

Notice that $e(G, H) = 0$ if and only if H has finite index in G. If H has infinite index in G then we have

$$e(G, H) = 1 + \dim \mathrm{Ker}\left(H^1(G, \mathcal{F}(H\backslash G)) \longrightarrow H^1(G, \mathcal{P}(H\backslash G))\right). \tag{4.3}$$

This can be seen by applying the long exact sequence of cohomology to the short exact sequence $\mathcal{F}(H\backslash G) \rightarrowtail \mathcal{P}(H\backslash G) \twoheadrightarrow \mathcal{P}(H\backslash G)/\mathcal{F}(H\backslash G)$.

A subset S of G is called H-finite if it is contained in a finite union of cosets of H; or more precisely if there is a finite subset F of G such that $S \subseteq HF$. The set of all H-finite subsets of G is denoted by $\mathcal{F}_HG$. We can view $\mathcal{F}_HG$ as an $\mathbb{F}_2G$-submodule of $\mathcal{P}G$, and now the second end invariant can be defined:

$$\tilde{e}(G, H) := \dim(\mathcal{P}G/\mathcal{F}_HG)^G. \tag{4.4}$$

As with the first, $\tilde{e}(G, H) = 0$ if and only if H has finite index in G. When H has infinite index a simple cohomology argument yields

$$\tilde{e}(G, H) = 1 + \dim H^1(G, \mathcal{F}_HG). \tag{4.5}$$

Further properties of e can be found in Scott's paper [21]. The invariant $\tilde{e}$ was introduced in [13] where many of its properties are established. In general, e is the more sensitive invariant: in geometric settings it carries more delicate information than $\tilde{e}$ and consequently one might expect it to be more useful. However, it turns out that the very insensitivity of $\tilde{e}$ can be a great advantage. There are three main reasons why $\tilde{e}$ is so useful. First, the formula (4.5) is a good deal simpler than the corresponding formula (4.3) for e and as a consequence $\tilde{e}$ is often much easier to compute than e. As an example, we recall the following facts from [13].

Lemma 4.6. *Let G be a PD^n-group and let H be a PD^{n-1}-subgroup. Then*

(i) $\tilde{e}(G, H) = 2$ and $e(G, H)$ is equal to 1 or 2,

(ii) $e(G, H) = 2$ if and only if the restriction to H of the dualising module for G is isomorphic to the dualising module for H.

Two further properties which make $\tilde{e}$ useful are the following

Lemma 4.7. $e(G, H) \leqslant \tilde{e}(G, H)$.

Lemma 4.8. *If $H \leqslant K$ are subgroups of infinite index in G then $\tilde{e}(G, H) \leqslant \tilde{e}(G, K)$.*

Both of these facts are easy to prove and we refer the reader to [13] for details. Their significance arises from the fact that end invariants are particularly interesting when they take values $\geqslant 2$. Lemma 4.7 suggests that $\tilde{e}(G, H)$ is more likely to be at least 2 than $e(G, H)$. Lemma 4.8 provides enormous room for manoeuvre: once we know that $\tilde{e}(G, H) \geqslant 2$ for some subgroup H we can enlarge H with the sole proviso that we should not enlarge it to a subgroup of finite index. These considerations turn out to be very important in proving Theorem 1.2.

What kind of group theoretic results can one hope to prove when given $e(G, H) \geqslant 2$ or $\tilde{e}(G, H) \geqslant 2$? Since the first condition is stronger than the second we expect to be able to draw stronger conclusions from the first. This is indeed true, and is nicely illustrated by the following theorem which we prove in this section. For convenience we shall say that a group G splits over a subgroup H if and only if either $G = K *_H L$ is a non-trivial amalgamated free product or $G = B*_{H,t}$ is an HNN-extension, where H is the amalgamated or associated subgroup in each case. We say that H is a malnormal subgroup of G if and only if $H \cap H^g = 1$ for all $g \notin H$.

Theorem 4.9. *Let G be a group and let H be a proper malnormal subgroup. Assume that $e(G) = e(H) = 1$. Then*

(i) G splits over H if and only if $e(G, H) \geqslant 2$; and

(ii) G splits over a subgroup of H if and only if $\tilde{e}(G, H) \geqslant 2$.

The assumptions that $e(G) = e(H) = 1$, which refer to the ordinary number of ends rather than ends of pairs, should not be regarded as serious restrictions: one can view Stallings' classical theorem on groups with more than one end as an assertion that most infinite groups have one end. On the other hand, the assumption that H is malnormal is a very serious restriction, but it turns out to be satisfied at a critical step in the proof of Theorem 1.2. It is worth noting two examples which show that one cannot drop either of the conditions $e(H) = 1$ or the malnormality of H. Notice that both examples pertain to low dimensional manifold theory.

Example 4.10. *Let G be the fundamental group of a closed orientable surface of genus at least 2 and let H be an isolated infinite cyclic subgroup. Then $e(G) = 1$, H is malnormal and $e(G, H) = \tilde{e}(G, H) = 2$. However, such an H can be chosen so that G does not split over H or any subgroup of H.*

In this example we have $e(H) = 2$, which is why Theorem 4.9 does not apply. Suitable choices of H can be made by choosing as a generator an element of G which is not a proper power and which corresponds to a loop on the surface which is not freely homotopic to an embedded circle.

Example 4.11. *Let G be the group with presentation*

$$\langle x, y, z; x^2 = y^3 = z^7 = xyz \rangle,$$

and let H be any rank 2 free abelian subgroup. Then $e(G) = e(H) = 1$ and $e(G, H) = \tilde{e}(G, H) = 2$, but G does not split over any subgroup whatsoever.

Here, G is the fundamental group of a closed Seifert fibered 3-manifold, and it contains an abundance of free abelian subgroups of rank 2. However, every such subgroup meets the centre of G and so is not malnormal. In both 4.10 and 4.11, the assertions $e(G, H) = \tilde{e}(G, H) = 2$ follow directly from Lemma 4.6.

In one direction, Theorem 4.9 is straightforward for it is an elementary fact that if G splits over H then $e(G, H) \geqslant 2$. If G splits over a subgroup H_0 of H then $e(G, H_0) \geqslant 2$ and it follows from Lemmas 4.7 and 4.8 that $\tilde{e}(G, H) \geqslant 2$ in this case. We shall concentrate on the more difficult implication for (ii) since (i) is similar but easier. In fact Theorem 4.9 is a very special case of results proved in [15], but we can dramatically simplify the proof given there. We shall need some further notation and some preliminary Lemmas. We write $A + B$ for the symmetric difference of two subsets A and B of G, and we denote the complement of B in G by B^*. A subset B is called H-almost invariant if and only if $B + Bg$ is H-finite for all $g \in G$. We shall say that B is a proper H-almost invariant subset if it is H-almost invariant and in addition neither B nor B^* is H-finite. The next Lemma is merely a translation of the definition of $\tilde{e}$ into this new notation.

Lemma 4.12. *Let $H \leqslant G$ be groups and suppose that $\tilde{e}(G, H) \geqslant 2$. Then G has a proper H-almost invariant subset.*

Lemma 4.13. *If H is a malnormal subgroup of G and $e(H) = 1$ then $H^1(H, \mathcal{F}_H G)$ is zero.*

Proof. As an $\mathbb{F}_2 G$-module, $\mathcal{F}_H G$ is isomorphic to the induced module

$$\mathrm{Ind}_H^G \mathcal{P} H.$$

Mackey decomposition gives its structure as an $\mathbb{F}_2 H$-module:

$$\operatorname{Res}_H^G \operatorname{Ind}_H^G \mathcal{P}H \cong \bigoplus_g \operatorname{Ind}_{H \cap H^g}^H \mathcal{P}Hg.$$

But since H is malnormal, this simplifies at once to show that, as an $\mathbb{F}_2 H$-module, $\mathcal{F}_H G$ is a direct sum of $\mathcal{P}H$ and a free $\mathbb{F}_2 H$-module. Now the Lemma follows because $H^1(H, \mathcal{P}H)$ vanishes for any group H whatever, and $H^1(H, -)$ vanishes on free modules for any group H with one end.

Taking Lemmas 4.12 and 4.13 together one can deduce

Corollary 4.14. *Let $H \leqslant G$ be groups with $\tilde{e}(G, H) \geqslant 2$, $e(H) = 1$ and H malnormal. Then there is a proper H-almost invariant subset B of G such that $B = BH$.*

Proof. Any H-almost invariant set B corresponds to a derivation from G to $\mathcal{F}_H G$, given by $g \mapsto B + Bg$. Since $H^1(H, \mathcal{F}_H G) = 0$ it follows that the restriction of this derivation to H is inner. This means that there is an H-finite set C such that for all $h \in H$, we have $B + Bh = C + Ch$. Now replace B by $B + C$.

Now we come to the core of the argument. The basic principle was discovered by Stallings. In order to obtain a splitting of a group from a given H-almost invariant subset one would like to find a bipolar structure in the group. After Stallings' original work [26], Dunwoody discovered an elegant way of viewing this via actions on trees. These methods are described in further detail here, in Roller's article, [19]. The main point is to try to prove that

For all $g \in G$ one of the sets $gB \cap B, gB \cap B^*, gB^* \cap B, gB^* \cap B^*$ is empty.

(4.15)

More precisely, we shall obtain splittings of groups by using the following form of the Stallings-Dunwoody theory:

Theorem 4.16. *Let K be a subgroup of G and let B be a proper K-almost invariant set which is contained in $K^* = G \smallsetminus K$. Suppose that the following hold.*

(i) $gB^* \cap B^* = \varnothing$ for all $g \in B \cap B^{-1}$.

(ii) $gB \cap B^* = \varnothing$ for all $g \in B \cap B^{*-1}$.

(iii) $gB^* \cap B = \varnothing$ for all $g \in B^* \cap B^{-1}$.

(iv) $gB \cap B = \varnothing$ for all $g \in B^* \cap B^{*-1} \smallsetminus K$.

(v) $gB = B$ for all $g \in K$.

(vi) $gB \neq B^*$ for all $g \in G$.

Then G splits over K.

We refer the reader to Roller's article [19] for details of how this can be proved.

Ensuring that the first four conditions of Theorem 4.16 are satisfied in a particular situation can be difficult, and can require modifying the original choice of almost invariant subset. In Stallings original work on groups with infinitely many ends, this was undoubtedly a major problem, and it was remarkable that he could overcome it. However, in proving Theorem 4.9 things are somewhat easier, because of the very strong condition that H is malnormal. The next Lemma is the key to making effective use of the malnormal condition.

Lemma 4.17. *Let H be a subgroup of G and let $A = AH$ and $B = BH$ be H-almost invariant subsets of G. If g belongs to $A^{*-1} \cap B^*$ then $gA \cap B$ is $(H^{g^{-1}} \cap H)$-almost invariant.*

Proof. Let x be any element of G. We must show that the symmetric difference $(gA \cap B)x + (gA \cap B)$ is $(H^{g^{-1}} \cap H)$-finite. Since A and B are H-almost invariant, there exist finite sets E, F such that $A + Ax \subseteq HE$ and $B + Bx \subseteq HF$. Now we have

$$(gA \cap B)x + (gA \cap B) = (gAx + gA) \cap Bx + gA \cap (Bx + B)$$
$$\subseteq gHE \cap Bx + gA \cap HF$$
$$= gHE \cap Bx + g(A \cap g^{-1}HF).$$

Now $gHE \cap Bx$ is plainly $H^{g^{-1}}$-finite, but it is also H-finite because gH is contained in B^* and E is finite. Now a set which is both S-finite and T-finite for two subgroups S and T is automatically $S \cap T$-finite. Thus $gHE \cap Bx$ is $(H^{g^{-1}} \cap H)$-finite. Similarly, using the fact that g^{-1} belongs to A^*, it follows that $A \cap g^{-1}HF$ is $H \cap H^g$-finite and hence $g(A \cap g^{-1}HF)$ is $(H^{g^{-1}} \cap H)$-finite. This proves the Lemma.

We need one further Lemma in order to prove Theorem 4.9.

Lemma 4.18. *Let $H \leqslant G$ be groups and suppose that there is at least one element $g \in G$ such that the left coset gH is not H-finite. Suppose that $A = AH$ is a proper H-almost invariant subset of G. Then there is a subset B of $G \smallsetminus H$ with the following properties.*

(i) For all $h \in H$, either $hB = B$ or $hB \cap B = \varnothing$.

(ii) B is a proper K-almost invariant subset, where $K = \{h \in H ; hB = B\}$.

(iii) $B = BH$.

Proof. Replacing A by A^* if necessary, we may assume that g belongs to A. Since A is H-almost invariant, there is a finite set F_x, for each $x \in G$ such that $A + Ax \subseteq HF_x$. Let $\mathfrak{Y}$ denote the set of all those subsets C of G which satisfy the conditions (a)–(c) below.

(a) $C + Cx \subseteq HF_x$ for all $x \in G$.

(b) $C = CH$.

(c) $gH \subseteq C$.

Note that $\mathfrak{Y}$ is non-empty because A is in it. Let B be the intersection of all the members of $\mathfrak{Y}$. It is easy to check that $B \in \mathfrak{Y}$, and so in fact B is the unique smallest member of $\mathfrak{Y}$. Moreover, (a) guarantees that B is H-almost invariant, and (c) guarantees that B is not H-finite. Since $B \subseteq A$ holds as well we see that B is a proper H almost invariant subset.

If h is an element of H then hB satisfies both (a) and (b) above, so hB belongs to $\mathfrak{Y}$ if and only if (c) holds. But if hB is in $\mathfrak{Y}$ then $B \subseteq hB$ because B is the smallest member of $\mathfrak{Y}$. Suppose, if possible that B is a proper subset of hB. Then $h^{-1}B$ is a proper subset of B and by the same token, since it satisfies (a) and (b), we conclude that g does not belong to $h^{-1}B$, in which case g does belong to $B \smallsetminus h^{-1}B$. But it is easy to check that (a) and (b) hold for $B \smallsetminus h^{-1}B$, and so we have found a member of $\mathfrak{Y}$ which is smaller than B, a contradiction. Therefore $B \subseteq hB$ implies $B = hB$. On the other hand, if g does not belong to hB then $B \smallsetminus hB$ belongs to $\mathfrak{Y}$ and so $B \subseteq B \smallsetminus hB$ and $B \cap hB = \varnothing$. Thus B satisfies (i) and (iii) of the Lemma.

Finally, set $K := \{h \in H \; ; \; hB = B\}$. The fact that B satisfies (i) ensures that for any x in G, $B \cap Hx$ consists of at most one coset of K. This, together with the fact that B is H-almost invariant, shows that B is K-almost invariant.

The Proof of Theorem 4.9. We begin by proving the harder part (ii). Let G and H be as in the statement of the Theorem and suppose that $\tilde{e}(G, H) \geqslant 2$. By Corollary 4.14 there exists a proper H-almost invariant subset A such that $A = AH$. Now, if g is any element of $G \smallsetminus H$ then gH is not H-finite: after all, gH is certainly $H^{g^{-1}}$-finite and if it were H-finite as well then it would be $H \cap H^{g^{-1}}$-finite or in fact *finite* because H is malnormal; a contradiction. Thus the special hypothesis of Lemma 4.18 is satisfied and we can apply this Lemma to obtain a subgroup K of H and a proper H almost invariant set B such that $B = KBH$ and $B \cap hB = \varnothing$ for all $h \in H \smallsetminus K$.

The next step is to check that B satisfies the hypotheses (i)–(iv) of Theorem 4.16. Suppose that g belongs to $B^* \cap B^{*-1} \smallsetminus H$. Applying Lemma 4.17 with $A = B$, we deduce that $gB \cap B$ is $H^{g^{-1}} \cap H$-almost invariant. But the malnormality of H implies that $H^{g^{-1}} \cap H = 1$ and therefore $gB \cap B$ is almost invariant in the classical sense. Since G has one end, it has no proper almost invariant subsets, and it follows that $gB \cap B$ is finite. Now note that $gB \cap B$ is right invariant under the infinite group H and being finite, it must be empty. This establishes the hypothesis 4.16(iv) for those group elements in $B^* \cap B^{*-1} \smallsetminus H$. But we have already seen that it holds for $g \in H \smallsetminus K$ because this is part of the conclusion of Lemma 4.18. Hence 4.16(iv) is established. Checking (i)–(iii) of 4.16 is carried out in just the same way, by applying Lemma 4.17 with each of A and B replaced by one of B and B^*.

We know that $B = KB$. Since B is a proper K-almost invariant set, it is easy to see that $gB \neq B$ for all $g \notin K$. Therefore there is no g for which

$gB = B^*$, for such a g must normalize K and therefore must belong to H, a contradiction. Thus a splitting of G over the subgroup K of H follows from Theorem 4.16.

Proving Theorem 4.9(i) is similar but simpler because if $e(G, H) \geqslant 2$ we can choose B to satisfy $B = HBH$ and there is no need to appeal to Lemma 4.18. The proof now proceeds to yield a splitting of G over H.

5. The proof of Theorem 1.2

We begin with a simple remark:

Lemma 5.1. *If G is a PD^n-group and G splits over a subgroup H then H has cohomological dimension exactly $n - 1$.*

Proof. Certainly H and the vertex groups involved in the splitting must have infinite index in G and therefore both H and the vertex groups have cohomological dimension at most $n - 1$ by Strebel's Theorem 3.2. Now, corresponding to the splitting of G there is a Mayer-Vietoris sequence for the cohomology of G and this ends as follows:

$$\cdots \longrightarrow H^{n-1}(H, -) \longrightarrow H^n(G, -) \longrightarrow 0.$$

Applying this with coefficient module $\mathbb{Z}G$ we conclude that $H^{n-1}(H, \mathbb{Z}G)$ is non-zero, because $H^n(G, \mathbb{Z}G) = \mathbb{Z}$ is non-zero. This shows that H also has cohomological dimension at least $n - 1$.

Now suppose that G is a PD^3-group and that H is a rank 2 free abelian subgroup. Then $\tilde{e}(G, H) = 2$, as noted in Lemma 4.6, and so if H is malnormal in G then G splits over a subgroup K of H by Theorem 4.9. Note that K must necessarily have finite index in H because of Lemma 5.1. What this remark suggests is that it is important to understand how H intersects its conjugates. Malnormality is a strong condition which does not hold in general, but clearly, for any g in G, one of the following holds:

$H \cap H^g = 1$, or

$H \cap H^g$ is infinite cyclic, or

$H \cap H^g$ has finite index in both H and H^g.

Group elements g for which the third possibility holds are precisely the elements for which H and H^g are commensurable and the set of all such elements is a subgroup called the commensurator of H in G, and denoted by $\mathrm{Comm}_G(H)$. In [10] it is shown that either H has finite index in its commensurator or G is polycyclic-by-finite, and at this point the proof naturally divides into two cases. If G is polycyclic-by-finite then it is easy to unravel the structure of G and we shall say no more about this case. Thus we assume from now on that H has finite index in its commensurator. Since G is torsion-free there are only two possible groups which $\mathrm{Comm}_G(H)$ can be, up to isomorphism; the free abelian group of rank 2 or the Klein bottle group. Replacing H by its commensurator, we still have $\tilde{e}(G, H) = 2$ but now we may assume that $H = \mathrm{Comm}_G(H)$ and hence that for all $g \notin H$, either

$H \cap H^g = 1$, or

$H \cap H^g$ is infinite cyclic.

Thus, in view of Theorem 4.9 we have

Proposition 5.2. *If G does not split over a subgroup of finite index in H then there exists an element $g \in G$ such that $H \cap H^g$ is infinite cyclic.*

In order to see why this Proposition is important we need a more sophisticated version of Theorem 4.9(ii). This version has the advantage that the malnormal assumption on H is considerably weakened (and replaced by the two assumptions that $\tilde{e}(G, H \cap H^g) = 1$ for all $g \in G \smallsetminus H$ and that there is a group element g such that gH is not H-finite), but at the expense of assuming that there exists an H-almost invariant set which satisfies the conclusion of Corollary 4.14.

Theorem 5.3. *Let $H \leqslant G$ be groups and suppose that there is a proper H-almost invariant subset B of G such that $B = BH$. Assume that $e(G) = 1$, and that $\tilde{e}(G, H \cap H^g) = 1$ for all $g \in G \smallsetminus H$. Then G splits over a subgroup of H.*

Outline of proof. As in the proof of Theorem 4.9 one finds a subgroup K of G and a new H-almost invariant subset which now replaces B, such that $B = KBH$ and $hB \cap B = \varnothing$ for all $h \in H \smallsetminus K$. The important thing now is to check that the hypotheses (i)–(iv) of Theorem 4.16 hold. When checking (iv) for example, one finds that for g belonging to $B^* \cap B^{*-1} \smallsetminus H$, the set $gB \cap B$ is $H^{g^{-1}} \cap H$-almost invariant. Since $\tilde{e}(G, H^{g^{-1}} \cap H) = 1$ it follows that $gB \cap B$ is $H^{g^{-1}} \cap H$-finite. If $gB \cap B$ is non-empty then being right invariant under H means that there is an x such that xH is $H^{g^{-1}} \cap H$-finite, and this leads to a contradiction. Therefore $gB \cap B$ is empty, and the rest of the proof proceeds as with the proof of 4.9.

Returning now to the PD^3-group G and the self-commensurating torus or Klein bottle subgroup H, we note that all the hypotheses of Theorem 5.3 are satisfied except one: namely we do not know that there is a proper H-almost invariant subset B which is equal to BH. Certainly there exist proper H-almost invariant subsets, but the question of whether or not they can be chosen to satisfy $B = BH$ is a genuine obstruction to splitting G over a subgroup of H. This fact, and the theory of this obstruction can be found in in [10].

To exploit this we will use the following technical result which replaces the idea of singularity controllers introduced in [12].

Proposition 5.4. *Let $H \leqslant G$ be groups such that H has finite index in $\mathrm{Comm}_G(H)$ and suppose that B is a proper H-almost invariant subset of G. Let K and K_1 be finitely generated normal subgroups of H, and let N and N_1 denote their normalizers in G. Suppose that the following conditions hold.*

(i) $B = HB$.

(ii) For all non-empty finite subsets F of G, HFK is a proper subset of HFN and HFK_1 is a proper subset of HFN_1.

(iii) For all $g \in G$, H^g is H-finite if and only if $H^{g^{-1}}$ is H-finite.

(iv) Whenever L and L_1 are subgroups of finite index in K and K_1 respectively then the group $\langle L, L_1 \rangle$ which they generate has finite index in H.

Then $B_0 = B \smallsetminus \mathrm{Comm}_G(H)$ satisfies $B_0 = B_0 H$.

Proof. Initially we will not use the subgroups K_1 and N_1, nor will we need condition (iv), so that the first part of the argument can be applied in slightly more general situations. First, we would like to know that $B = BK$. While this may not be true, it does at least follow from the fact that K is finitely generated that there are only finitely many double cosets HtK with the property that HtK is not contained in B and is also not contained in B^*. We now replace B by another set which differs from B by an H-finite amount in order to minimize the number of these bad double cosets. Let $Ht_1 K, \ldots, Ht_m K$ be a list of the bad double cosets. These now have the property that neither $Ht_i K \cap B$ nor $Ht_i K \cap B^*$ is H-finite, for each i, whereas all other double cosets are either entirely contained in B or entirely contained in B^*.

Set $C := \{g \in G \; ; \; gK$ is H-finite$\}$, (note that in general this need not be a subgroup). Clearly $t_1, \ldots, t_m$ do not belong to C. Now let g be an element of N. Note that $N \subseteq C$ and $C = HCN$, so the H finite set $B + Bg$ is the disjoint union of the sets $(B \cap C) + (B \cap C)g$ and $(B \smallsetminus C) + (B \smallsetminus C)g$, and furthermore $B + Bg$ is contained in the disjoint union of C and the double cosets $Ht_i K$, for $1 \leqslant i \leqslant m$. Let T be a set of (H, K)-double coset representatives in G, let T_0 consist of those elements of T which belong to C, let T_1 consist of those elements of T which belong to $G \smallsetminus (C \cup \{t_1, \ldots, t_m\})$ and let T_2 be equal to $\{t_1, \ldots, t_m\}$. Thus T is the disjoint union of T_0, T_1 and T_2. Moreover we can choose subsets S_0 and S_1 of T_0 and T_1 respectively such that B is the disjoint union of $HS_0 K$, $HS_1 K$ and the sets $B \cap Ht_i K$ for $1 \leqslant i \leqslant m$. The fact that $B + Bg$ is H-finite for all $g \in N$ shows that $HS_1 K = HS_1 N$ and that the double cosets $Ht_i K$ are permuted under right multiplication by elements $g \in N$. Hence $HT_2 K = HT_2 N$ and so by (ii), it follows that $T_2 = \varnothing$. Thus we have proved that

$$B = BK \text{ and } (B \smallsetminus C) = (B \smallsetminus C)N. \tag{5.5}$$

Now we introduce K_1 and we set $C_1 = \{g \in G \; ; \; gK_1$ is H-finite$\}$. By exactly the same reasoning,

$$(B \smallsetminus C_1) = (B \smallsetminus C_1)N_1. \tag{5.6}$$

Now H is contained in $N \cap N_1$, so both $B \smallsetminus C$ and $B \smallsetminus C_1$ are right invariant under H, and therefore their union is right invariant under H:

$$(B \smallsetminus (C \cap C_1))H = (B \smallsetminus (C \cap C_1))H. \qquad (5.7)$$

Claim. $C \cap C_1$ *is H-finite.*

Proof of Claim. Let g be an element of $C \cap C_1$. Then gK and gK_1 are both H-finite. Hence $K^{g^{-1}}$ and $K_1^{g^{-1}}$ are both H-finite. This implies that $L = K \cap H^g$ has finite index in K and $L_1 = K_1 \cap H^g$ has finite index in K_1. Therefore $\langle L, L_1 \rangle$ has finite index in H by (iv), but it is also contained in H^g. This implies that H^g is H-finite and hence by (iii), g belongs to $\mathrm{Comm}_G(H)$. We conclude that $C \cap C_1 \subseteq \mathrm{Comm}_G(H)$. Since H has finite index in $\mathrm{Comm}_G(H)$ the claim is proved.

Now the Proposition clearly follows from (5.7).

Corollary 5.8. *Let G be a PD^3-group which is not polycyclic and let H be a rank 2 free abelian subgroup such that $e(G, H) = 2$ and such that H has finite index in its commensurator. If G does not split over a subgroup commensurable with H then there is a cyclic subgroup K of H such that for all $g \in G$,*

$$H \cap H^g \text{ infinite cyclic} \Rightarrow H \cap H^g \text{ is commensurable with } K.$$

Moreover $C := \{g \in G \; ; \; gK \text{ is } H\text{-finite}\}$ is equal to the commensurator of K.

Proof. There must be at least one infinite cyclic subgroup K of the form $H \cap H^g$ for some g by Proposition 5.2. If there is another one, K_1, which is not commensurable with K then it is easy to see that all the hypotheses of Proposition 5.4 are satisfied. The conditions in 5.4(ii) come about because $K = H \cap H^g$ necessarily has very large normalizer, but the proof is quite technical and it is here that we use the assumtion that G is not polycyclic. First note that K is central in the group $\langle H, H^g \rangle$ because H is abelian. There are two cases to consider. If $\langle H, H^g \rangle$ has finite index in G then K and in particular every element of K has centraliser of finite index in G. This means that the set

$$\Delta(G) := \{x \in G \; ; \; |G : C_G(x)| < \infty\}$$

is non-trivial. Standard results show that, in any torsion-free group, $\Delta(G)$ is a torsion-free abelian normal subgroup, (details can be found in [18]). If $\Delta(G)$ has rank $\geqslant 2$ then it is straightforward to show that G must be polycyclic. Therefore $\Delta(G)$ has rank one and hence every cyclic subgroup of $\Delta(G)$ is normal. In particular K is normal and so in this case the Corollary is established.

The second case we have to consider is when $\langle H, H^g \rangle$ has infinite index in G. In this case it must be a group of cohomological dimension 2 and since it has the non-trivial central subgroup K it follows from results of Bieri [1] that $\langle H, H^g \rangle / K$ is free-by-finite, and since $\langle H, H^g \rangle / K$ cannot be cyclic-by-finite it follows that $\langle H, H^g \rangle$ involves a non-cyclic free subgroup. It is now easy to check the hypothesis (5.4)(ii). Let N denote the normalizer of K and suppose that F is a non-empty finite subset of G such that $HFN = HFK$. Choose any element $x \in F$. Then $xN \subseteq HFK$ and so we have $N \subseteq x^{-1}HFK$. Now apply Dedekind's Law in the following form: if $A \leqslant B$ are subgroups and C is a subset of some group then $B \cap CA = (B \cap C)A$. Set $A := K$, $B := N$, and $C := x^{-1}HF$. Thus we have $N = N \cap x^{-1}HFK = (N \cap x^{-1}HF)K$. Now, $x^{-1}HF = H^x x^{-1}F$ comprises finitely many cosets of H^x, and an intersection $N \cap H^x y$ is either empty or consists of a single coset of $N \cap H^x$. Therefore there is a finite subset E of N such that $N = (N \cap x^{-1}HF)K = (N \cap H^x)EK$, and since the elements of E normalize K, we have $N = (N \cap H^x)KE$. Since $N \cap H^x$ is abelian and normalizes K, it follows that $(N \cap H^x)K$ is a metabelian subgroup of G. But then N is metabelian-by-finite, which contradicts the fact that it has a non-cyclic free subgroup.

The remaining hypotheses of Proposition (5.4) are relatively easy to check, and the Proposition shows that the obstruction to a splitting vanishes, a contradiction. The last assertion now follows easily.

The Proof of Theorem 1.2. Let G be a PD^3-group and let H be a rank 2 free abelian subgroup such that $e(G, H) = 2$ and such that H has finite index in its commensurator. Let $B = HB$ be a proper H-almost invariant subset of G. Let K be a cyclic subgroup of H as determined by Corollary 5.8. Let C denote the commensurator of K and let N denote the normalizer of K.

Now the first part of the proof of Proposition 5.4 shows that $(B \smallsetminus C)N = (B \smallsetminus C)$, (see (5.5)), and also after possible adjustment to B, that $BK = B$. Let t be an element of $B \smallsetminus C$ and let g be an element of C. Then HtK is contained in B, and since B is H-almost invariant, $HtKg \smallsetminus B$ is H-finite. Since K^g and K are commensurable we can write $HtKg = HtgK^g$ as a union of finitely many translates of $HtgL$ where $L = K \cap K^g$. Now $HtgL \smallsetminus B$ is H-finite and so also is any translate. Therefore $HtgK$, which is a union of finitely many translates, is H-finite. Since $B = BK$ we must have $HtgK \subseteq B \smallsetminus C$, and hence $HtC \subseteq B \smallsetminus C$ for all $t \in B \smallsetminus C$. Hence

$$(B \smallsetminus C)C = (B \smallsetminus C). \tag{5.9}$$

A fortiori, B is C-almost invariant and so also is $B_1 = B \smallsetminus C$. We are now almost in a position to apply Theorem 5.3, with C playing the role of H. But we need to consider some special cases first.

Case 1. $C = G$.

The function

$$\phi : g \mapsto \frac{|K^g : K \cap K^g|}{|K : K \cap K^g|}$$

defines a homomorphism from C to the multiplicative group of positive rational numbers. [To see that ϕ is a homomorphism note first that if L is any subgroup of finite index in $K \cap K^g$ then

$$\phi(g) = \frac{|K^g : L|}{|K : L|}.$$

Now, for elements $g, h \in G$, set $L := K \cap K^g \cap K^h \cap K^{gh}$ and observe that

$$\phi(g) = \frac{|K^g : K \cap K^g|}{|K : K \cap K^g|} = \frac{|K^{gh} : K^h \cap K^{gh}|}{|K : K \cap K^g|},$$

$$\phi(h) = \frac{|K^h : L|}{|K : L|} = \frac{|K^h : K^h \cap K^{gh}| \cdot |K^h \cap K^{gh} : L|}{|K : K \cap K^g| \cdot |K \cap K^g : L|} = \frac{|K^h \cap K^{gh} : L|}{|K \cap K^g : L|},$$

$$\phi(gh) = \frac{|K^{gh} : L|}{|K : L|} = \left(\frac{|K^{gh} : K^h \cap K^{gh}|}{|K : K \cap K^g|}\right) \cdot \left(\frac{|K^h \cap K^{gh} : L|}{|K \cap K^g : L|}\right) = \phi(g)\phi(h).]$$

Thus $G/\mathrm{Ker}\,\phi$ is free abelian. If $\mathrm{Ker}\,\phi$ is a proper subgroup of G then G has an infinite cyclic quotient and this leads at once to a splitting of G of HNN-type, fulfilling Theorem 1.2(ii). On the other hand, if trivial then for any $g \in G$, $K \cap K^g$ and all its subgroups are normalised by g. Thus if $g_1, \ldots g_n$ are generators of G then $K \cap K^{g_1} \cap \cdots \cap K^{g_n}$ is an infinite cyclic normal subgroup of G. Thus Theorem 1.2(iii) is fulfilled.

Case 2. *There exists $g \in G \smallsetminus C$ such that $K^g \cap C$ has finite index in K^g and $K \cap C^g$ has finite index in K.*

This means that there are non-identity elements x, y of K, K^g respectively such that $x \in \mathrm{Comm}_G(K^g)$ and $y \in \mathrm{Comm}_G(K)$. Thus there are non-zero integers p, q, r, s such that

$$(x^p)^y = x^q$$

and

$$(y^r)^x = y^s.$$

We can now define two homomorphisms ϕ_x and ϕ_y from $\langle x, y \rangle$ to $\mathbb{Q}^\times$ by $\phi_x(g) = \frac{a}{b}$ where a, b are such that $(x^b)^g = x^a$ and ϕ_y is defined by $\phi_y(g) = \frac{a}{b}$ where a, b are such that $(y^b)^g = y^a$. Using these homomorphisms it can be shown that there exist positive integers c, d such that x^c and y^d generate a free abelian group, A say, of rank 2. But now $\langle x^c \rangle$ and $\langle y^d \rangle$ are non-commensurable cyclic subgroups of A both having large commensurators, and therefore G splits over a subgroup commensurable with A by Corollary 5.8. Thus either (i) or (ii) of Theorem 1.2 is fulfilled.

Case 3. *C is a proper subgroup of G and for all $g \in G \setminus C$, either $K \cap C^g = 1$ or $K^g \cap C = 1$.*

In this case C must have infinite index and by Strebel's theorem it has cohomological dimension at most 2. We know that C contains a free abelian group H of rank 2 and since $\tilde{e}(G, H) = 2$ we have $\tilde{e}(G, C) \geqslant 2$. Moreover the finitely generated subgroups of C which meet K are $\mathfrak{X}$-groups and so are fundamental groups of finite graphs of infinite cyclic groups. From this structure theorem it follows that the subgroups of C which do not meet K are locally free. Hence for all $g \in G \setminus C$, the intersection $C \cap C^g$ is locally free, because either it can be regarded as a subgroup of C which does not meet K or it can be regarded as a subgroup of C^g which does not meet K^g. Now an easy calculation shows that if G is a PD^3-group and J is a locally free subgroup then $\tilde{e}(G, J) = 1$. Hence we have $\tilde{e}(G, C \cap C^g) = 1$ for all $g \notin C$. Thus, with cases 1 and 2 disposed of, we can return to the C-almost invariant set B_1. Theorem 5.3 now applies.

References

[1]   R. Bieri, *Homological dimension of discrete groups*, Queen Mary College, London, 1981.

[2]   A. Borel and J-P. Serre, *Corners and arithmetic groups*, Comment. Math. Helv. **48** (1973), pp. 436–491.

[3]   M. W. Davis, *Groups generated by reflections and aspherical manifolds not covered by Euclidean space*, Annals Math. **117** (1983), pp. 293–324.

[4]   Warren Dicks and M. J. Dunwoody, *Groups acting on graphs*, Cambridge University Press, 1989.

[5]   M. J. Dunwoody, *Accessibility and groups of cohomological dimension one*, Proc. London Math. Soc. (3) **38** (1979), pp. 193–215.

[6]   M. J. Dunwoody, *Cutting up graphs*, Combinatorica **2** (1982), pp. 15–23.

[7]   M. J. Dunwoody and M. A. Roller, *Splitting groups over polycyclic-by-finite subgroups*, to appear in Bull. London Math. Soc.

[8]   J. A. Hillman, *Seifert fibre spaces and Poincaré duality groups*, Math. Z. **190** (1985), pp. 365–369.

[9]   J. A. Hillman, *Three dimensional Poincaré duality groups which are extensions*, Math. Z. **195** (1987), pp. 89–92.

[10] P. H. Kropholler and M. A. Roller, *Splittings of Poincaré duality groups*, Math. Z. **197** (1988), pp. 421–438.

[11] P. H. Kropholler and M. A. Roller, *Splittings of Poincaré duality groups II*, J. London Math. Soc. **38** (1988), pp. 410–420.

[12] P. H. Kropholler and M. A. Roller, *Splittings of Poincaré duality groups III*, J. London Math. Soc. **39** (1989), pp. 271–284.

[13] P. H. Kropholler and M. A. Roller, *Relative ends and duality groups*, J. Pure Appl. Algebra **61** (1989), pp. 197–210.

[14] P. H. Kropholler, *A note on centrality in 3-manifold groups*, Math. Proc. Camb. Phil. Soc. **107** (1990), pp. 261–266.

[15] P. H. Kropholler, *An analogue of the torus decomposition theorem for certain Poincaré duality groups*, Proc. London Math. Soc. (3) **60** (1990), pp. 503–529.

[16] P. H. Kropholler, *Baumslag-Solitar groups and some other groups of cohomological dimension two*, Comment. Math. Helvetici **65** (1990), pp. 547–558.

[17] G. Mess, *Examples of Poincaré duality groups*, Proc. Amer. Math. Soc. **110** (1990), pp. 1145–1146.

[18] D. S. Passman, *The algebraic structure of group rings*, Wiley, New York, 1977.

[19] M. A. Roller, *Constructing Group Actions on Trees*, these Proceedings.

[20] P. Scott, *Finitely generated 3-manifold groups are finitely presented*, J. London Math. Soc. (2) **6** (1972), pp. 437–440.

[21] P. Scott, *Ends of pairs of groups*, J. Pure Appl. Algebra **11** (1977), pp. 179–198.

[22] P. Scott, *A new proof of the annulus and torus theorem*, Amer. J. Math. **102** (1980), pp. 241–277.

[23] P. Scott, *The geometries of 3-manifolds*, Bull. London Math. Soc. **15** (1983), pp. 401–487.

[24] P. Scott, *Strong annulus and torus theorems and the enclosing property of characteristic submanifolds of 3-manifolds*, Quart. J. Math. Oxford (2) **35** (1984), pp. 485–506.

[25] J-P. Serre, *Trees*, Springer, Berlin, 1980.

[26] J. R. Stallings, *On torsion-free groups with infinitely many ends*, Ann. Math. **88** (1968), pp. 312–334.

[27] R. Strebel, *A remark on subgroups of infinite index in Poincaré duality groups*, Comment. Math. Helv. **52** (1977), pp. 317–324.

[28] R. Swan, *Groups of cohomological dimension one*, J. Algebra **12** (1969), pp. 585–601.

[29] C. B. Thomas, *Splitting theorems for certain PD^3-groups*, Math. Z. **186** (1984), pp. 201–209.

$\mathcal{N}$-Torsion and Applications

Martin Lustig and Yoav Moriah

Mathematisches Institut, Ruhr-Universität Bochum, D 4630 Bochum 1, Germany.
Department of Mathematics, Technion, Haifa 32000, Israel.

0. Introduction

In this article we present a short survey of the torsion invariant $\mathcal{N}(G)$ and its application to problems in topology and group theory, as developed in [LM1], [LM2] and [LM3]. The invariant is designed to distinguish non Nielsen equivalent generating systems of minimal cardinality for finitely generated groups (see Definition 1.1 below). It can be applied in practice to any finitely presented group and sometimes also to finitely generated groups which are not finitely presented (see [MoS]). Its main application is given by the following:

Theorem 0.1. *Let G be presented by*

$$G = \langle x_1, \ldots, x_n \mid R_1, R_2, \ldots \rangle.$$

Let $y_1, \ldots, y_n$ be a second generating system of G, given as words in the x_i, i.e.,

$$y_j = W_j(x_1, \ldots, x_n), \qquad for\ j = 1, \ldots, n.$$

Let $\partial W_j / \partial x_i$ and $\partial R_k / \partial x_i$ denote the canonical image in $\mathbb{Z}G$ of the Fox derivatives of W_j and R_k with respect to x_i. Let A be a commutative ring with $0 \neq 1 \in A$, and let $\rho : \mathbb{Z}G \to \mathbb{M}_m(A)$, $\rho(1) = 1$, be a ring homomorphism satisfying $\rho(\partial R_k / \partial x_i) = 0$ for all R_k and x_i. Then

(1) *$x_1, \ldots, x_n$ and $y_1, \ldots, y_n$ are both generating systems of minimal cardinality for G.*

(2) *If the determinant of the $(mn \times mn)$-matrix $(\rho(\partial W_j / \partial x_i)_{j,i})$ is not contained in the subgroup of A^* generated by the determinants of $\rho(\pm x_1), \ldots, \rho(\pm x_n)$, then $x_1, \ldots, x_n$ and $y_1, \ldots, y_n$ are not Nielsen equivalent.*

Supported by a grant from the German-Israeli Foundation for Research and Developement (G.I.F.)

In §1 we review the basic definitions and properties of $\mathcal{N}$. In §2 we describe an application of Theorem 0.1 to a class of non-trivial examples (where G is an amalgamated free product with amalgam isomorphic to $\mathbb{Z} \oplus \mathbb{Z}$), and in §3 we show the topological relevance of these examples, which occur in connection with work of Morimoto-Sakuma (see [MSa]) on unknotting tunnels for non-simple knots.

1. $\mathcal{N}$-torsion

Every generating system $x = \{x_1, \ldots, x_n\}$ of a finitely generated group G canonically determines a free group $F(X)$ on the basis $X = \{X_1, \ldots, X_n\}$, and an epimorphism $\beta_x : F(X) \twoheadrightarrow G$, given by $\beta_x(X_i) = x_i$. Nielsen equivalence is an equivalence relation on generating systems for groups, which describes when two such maps β_x are essentially the same:

Definition 1.1. *The systems* $x = \{x_1, \ldots, x_n\}$ *and* $y = \{y_1, \ldots, y_n\}$ *are said to be* Nielsen equivalent *if there is an isomorphism* $\theta : F(Y) \to F(X)$ *such that* $\beta_x \theta = \beta_y$.

The reader should notice that, given x and y as above, there are always homomorphisms $\theta : F(Y) \to F(X)$ with $\beta_x \theta = \beta_y$. However, these homomorphisms are in general far from being isomorphisms. Note that, if θ is an isomorphism, then it defines an element $\tilde{\theta}$ of $Aut(F(X))$, given by $X_i \to \theta(Y_i)$. If $F(X)$ is a finitely generated free group then a well known theorem of Nielsen (see [MKS], pp. 162) states that the group $Aut(F(X))$ is generated by elementary automorphisms of two kinds:

(a) For some $i \in \{1, \ldots, n\}$ map $X_i \mapsto X_i^{-1}$, and $X_j \mapsto X_j$ for $j \neq i$.
(b) For some $i, j \in \{1, \ldots, n\}$, $i \neq j$, map $X_i \mapsto X_i X_j$, and $X_k \mapsto X_k$ for $k \neq i$.

(Notice that every permutation of the generators can be expressed as a product of isomorphisms of type (a) and (b): for a transposition this is an easy exercise.)

Denote by $\partial/\partial X_i : \mathbb{Z}F(X) \to \mathbb{Z}F(X)$ the i-th Fox derivative of the integer group ring $\mathbb{Z}F(X)$ (see e.g. [BZ], [LS], [F]). Any system of n words $W = \{W_1, \ldots, W_n\}$ in $F(X)$ defines a Jacobian $(\partial W_j/\partial X_i)$ over $\mathbb{Z}F$. In particular, if W is obtained from X by one application of an elementary automorphism of type (a) or (b), then the corresponding Jacobi matrices are

$$
\text{(a)} \;\; i \;\;
\begin{pmatrix}
1 & & & & \vdots & & & \\
& \ddots & & & & & 0 & \\
& & 1 & & & & & \\
\cdots & & & -X_i^{-1} & & & & \\
& & & & 1 & & & \\
& 0 & & & & \ddots & & \\
& & & & & & & 1
\end{pmatrix}
\quad \text{or} \quad \text{(b)} \;\;
\begin{matrix} \\ \\ i \\ \\ j \\ \\ \end{matrix}
\begin{pmatrix}
1 & & & \vdots & & \vdots & & \\
& \ddots & & & & & 0 & \\
\cdots & & 1 & \cdots & X_i & & & \\
& & & \ddots & & \vdots & & \\
\cdots & & & & 1 & & & \\
& 0 & & & & & \ddots & \\
& & & & & & & 1
\end{pmatrix},
$$

respectively.

The Fox derivatives satisfy the chain rule property (see [BZ], pp. 125): if $U = \{U_1, \ldots, U_n\}$ is another basis for $F(X)$, then for each $V \in \mathbb{Z}F(X)$ one has

$$
\frac{\partial V}{\partial X_i} = \sum_1^n \frac{\partial V}{\partial U_j} \cdot \frac{\partial U_j}{\partial X_i}.
$$

Hence, if there exists an isomorphism θ as in Definition 1.1, the automorphism $\tilde{\theta}$ of $F(X)$ is a product of elementary Nielsen automorphisms, and then the Jacobian $(\theta(Y_j)/\partial X_i) = (\tilde{\theta}(X_j)/\partial X_i)$ is a product of elementary matrices as above. In particular the matrix $\beta_x(\partial\theta(Y_j)/\partial X_i)$ is then a product of elementary matrices over $\mathbb{Z}G$.

Hence, showing that the last statement does not hold implies that the generating systems x and y are not Nielsen equivalent. This is the approach which we take here and which leads us to introduce $\mathcal{N}$-torsion.

The basic difficulties with this approach are as follows:

(a) In general we do not know how to define θ, as every element $y_i \in G$ determines the element $\theta(Y_i)$ only up to modifications within the coset $\beta_x^{-1}(y_i)$ with respect to $\ker \beta_x$.

(b) We need to decide whether the Jacobian $(\theta(Y_j)/\partial X_i)$ over the (in general non-commutative) ring $\mathbb{Z}G$ is a product of elementary matrices.

The first difficulty is overcome by considering the problem over a quotient ring $\mathbb{Z}G/I$ defined below, thus dividing out the non-uniqueness of θ. Notice that the Fox ideal I as defined below is the smallest kernel one needs to take for such purpose.

Definition 1.2. *Let $G = \langle x_1, \ldots, x_n \mid R_1, R_2, \ldots \rangle$ and $\beta_x : F(X) \to G$ be as above.*

(a) Let I_x be the two sided ideal in $\mathbb{Z}G$ generated by

$$
\{\beta_x(\partial R_k/\partial X_i) \mid k = 1, 2, \ldots, \; X_i \in X\},
$$

and let γ_x denote the quotient map $\mathbb{Z}G \twoheadrightarrow \mathbb{Z}G/I_x$.

(b) Define the Fox ideal of G to be the two sided ideal I in $\mathbb{Z}G$ generated by all I_x where x is a generating system of G with minimal cardinality. Denote the quotient map $\mathbb{Z}G \to \mathbb{Z}G/I$ by γ.

Remark 1.3.

*(1) An easy computation using the chain rule for Fox derivatives shows that
the ideal $I_x < \mathbb{Z}G$ is equal to the two sided ideal generated by*

$$\{\beta_x(\partial R/\partial X_i) \mid R \in \ker \beta_x, \ X_i \in X\},$$

*and hence independent of the choice of the relators R_k, which normally
generate the kernel of β_x.*

*(2) If X' is a different basis of $F(X)$ and x' its β_x-image in G then, using
the chain rule again, one obtains $I_x = I_{x'}$. Thus I_x is an invariant of
the Nielsen equivalence class of x.*

The Fox ideal has the following very useful properties (see [LM2], Lemma 1.3):

Lemma 1.4.   *Let $\theta_k : F(Y) \to F(X)$, $k = 1, 2$, be any two homomorphisms
which satisfy $\beta_x \circ \theta_k = \beta_y$. Then*

*(1) for any two minimal generating systems x, y of G the matrix $(\partial y/\partial x) \in
\mathbb{M}_n(\mathbb{Z}G/I)$, obtained from $\beta_x(\partial\theta k(Y_j)/\partial X_i)_{j,i}$ by applying the above map
$\gamma : \mathbb{Z}G \to \mathbb{Z}G/I$, is the same for $k = 1$ as for $k = 2$.*

(2) The matrix $(\partial y/\partial x)$ is invertible. $\qquad\qquad\qquad\qquad\qquad\square$

Hence, by considering the matrix $\beta_x(\partial\theta(Y_j)/\partial X_i)_{j,i}$ as a matrix over $\mathbb{Z}G/I$,
we can take "Fox derivatives in the group"; that is $\partial y_j/\partial x_i$ is well defined
regardless of the particular expression of y_j as a product of the x_i's.

In order to tackle the second difficulty we appeal to the standard method
for detecting elementary matrices over a given ring, provided by algebraic
K-theory.

Given a ring R denote the direct limit of $GL_n(R)$ for $n \to \infty$ by $GL(R)$. Let
$E(R)$ denote the subgroup of $GL(R)$ generated by the images of the elementary matrices with 1's on the diagonal and at most one non-zero entry $r \in R$
off the diagonal. Recall that $E(R)$ is the commutator subgroup of $GL(R)$
(see [M]), and that the first K-group is defined as $K_1(R) = GL(R)/E(R)$.
Let T be the image in $K_1(\mathbb{Z}G/I)$ of the subgroup generated by trivial units,
i.e. the matrices of the form (a) above, with elements $\pm g \in G$ on the diagonal.

Definition 1.5.

*(a) Assume $I \neq \mathbb{Z}G$. We define the torsion invariant $\mathcal{N}(G)$ as the following
quotient of the first K-group of the ring $\mathbb{Z}G/I$:*

$$\mathcal{N}(G) = K_1(\mathbb{Z}G/I)/T.$$

For any two minimal cardinality generating systems x, y of G define

$$\mathcal{N}(y, x) = [(\partial y/\partial x)] \in \mathcal{N}(G).$$

*(b) If $I = \mathbb{Z}G$ then we say that $\mathcal{N}(G)$ is degenerate and we formally define
$\mathcal{N}(G) = \{0\}$. In particular one has always $\mathcal{N}(y, x) = 0$.*

Since any two matrices which differ only by a product of elementary matrices
determine the same value in $\mathcal{N}(G)$, one obtains:

Proposition 1.6. $\mathcal{N}(y,x)$ *depends only on the Nielsen equivalence classes of x and y. If x and y are Nielsen equivalent then $\mathcal{N}(y,x) = 0 \in \mathcal{N}(G)$.*   □

Having now put the original question in a theoretically satisfying framework, our next problem is that any application of $\mathcal{N}(G)$ requires us to compute I, and this needs information about all Nielsen classes of G, which are in general unknown. This difficulty is dealt with by the next lemma (Lemma 2.1 of [LM2]).

Lemma 1.7. *Let x be a minimal generating system of G, and let A be any commutative ring with $0 \neq 1 \in A$. Every ring homomorphism $\sigma_x : \mathbb{Z}G/I_x \to \mathbb{M}_m(A)$ with $\sigma_x(1) = 1$ maps $\gamma_x(I)$ to $0 \in \mathbb{M}_m(A)$ and hence induces a ring homomorphism $\sigma : \mathbb{Z}G/I \to \mathbb{M}_m(A)$.*   □

All maps $\sigma : \mathbb{Z}G/I \to \mathbb{M}_m(A)$ induce, by the functoriality of K_1, a map

$$K_1(\sigma) : K_1(\mathbb{Z}G/I) \to K_1(\mathbb{M}_m(A)) = K_1(A),$$

(where the last equation is induced by "forgetting the brackets"). On $K_1(A)$ we have the determinant map $\det : K_1(A) \to A^*$ into the multiplicative group of units A^* of A. Let τ_σ denote the composition map

$$\tau_\sigma : GL(\mathbb{Z}G/I) \xrightarrow{\text{def}} K_1(\mathbb{Z}G/I) \xrightarrow{K_1(\sigma)} K_1(A) \xrightarrow{\det} A^*.$$

We define the subgroup T_σ of A^* to be the image $\tau_\sigma(T)$ of the set of trivial units $T \subset GL(\mathbb{Z}G/I)$ (see Definition 1.5). Summing up, we obtain the following proposition, which also proves Theorem 0.1 (2). Notice that statement (1) of Theorem 0.1 follows directly from the fact that for non-minimal generating systems the ideal I (or more precisely, its analogue,) is always equal to the whole group ring, (see [L], [LM3]).

Proposition 1.8. *Let x be a minimal generating system of G and A a commutative ring with $0 \neq 1 \in A$. Any representation $\sigma : \mathbb{Z}G/I \to \mathbb{M}_m(A)$ (or equivalently $\sigma_x : \mathbb{Z}G/I_x \to \mathbb{M}_m(A)$) with $\sigma(1) = 1$ induces a homomorphism*

$$\mathcal{N}(\sigma) : \mathcal{N}(G) \to A^*/T_\sigma .$$ □

A necessary condition for applying Proposition 1.8 (or Theorem 0.1) is to have a representation $\sigma : \mathbb{Z}G/I \to \mathbb{M}_m(A)$ with $\sigma(1) = 1$, where A is a commutative ring. It was shown in [LM2] that such representations do generally exist, and how one can find them. In the next section we construct such a representation for a particular class of groups, and apply Theorem 0.1.

Before finishing the section we should point out some of the properties of $\mathcal{N}(G)$ (see Theorem I of [LM2]):

Remark 1.9.

(1) For all $n \in \mathbb{N}$ the construction $\mathcal{N}$ describes a functor from the category $\mathbf{C}_n$, of groups with fixed rank n and surjective homomorphisms, to the category $\mathbf{Ab}$ of abelian groups. In particular we obtain for all objects G, H of $\mathbf{C}_n$ and any morphism $f : G \twoheadrightarrow H$

$$\mathcal{N}(f) : \mathcal{N}(y, x) \mapsto \mathcal{N}(f(y), f(x))$$

for all generating systems x and y of G with cardinality n.

(2) If $z = \{z_1, \ldots, z_n\}$ is another minimal generating system of G as above, then

$$\mathcal{N}(z, y) + \mathcal{N}(y, x) = \mathcal{N}(z, x).$$

2. An example

Consider the following amalgamated free product with amalgam isomorphic to $\mathbb{Z} \oplus \mathbb{Z}$:

$$G = \langle x, y \mid x^p y^{-q} \rangle \underset{\langle x^p = u, \; x^r y^{-1} = l \rangle}{*} \langle u, v \mid u^{-1} l^{-1} u l \rangle,$$

where

(1) $p, q, r \in \mathbb{Z}$ satisfy $p, q, r > 1$, $p - rq = 1$, and
(2) $l = l(v, u)$ is a word in v and u with exponent sum in v different from 0.

We can eliminate the generators y and u and obtain a two-generator, one-relator presentation:

$$G = \langle x, y \mid x^p (l^{-1} x^r)^{-q} \rangle, \qquad l = l(v, x^p).$$

Notice that since $x^p = y^q$ in the first factor, we can replace the amalgamation $x^p = u$ by $y^q = u$ without changing G:

$$G = \langle x, y \mid x^p y^{-q} \rangle \underset{\langle y^q = u, \; x^r y^{-1} = l \rangle}{*} \langle v, u \mid u^{-1} l^{-1} u l \rangle.$$

As $\gcd(p, r) = 1$, we can eliminate the generator x, applying the Euclidean algorithm to the equations $x^p = y^{-q}$, $x^r = ly$. Thus $\{y, v\}$ is a second generating system for G.

Proposition 2.1. *The generating systems $\{x, v\}$ and $\{y, v\}$ of G are not Nielsen equivalent.*

Proof. We want to apply Theorem 0.1. Hence our first step is to compute the canonical generators for the ideal $I_{\{x,v\}}$ from the presentation of G. Let $\beta : F(x, v) \twoheadrightarrow G$ be the natural epimorphism as above. As our group G is a two-generator, one-relator group, the generators of the ideal $I_{\{x,v\}}$ are:

$$\beta(\partial R/\partial v) = \beta(\partial x^p (l^{-1} x^r)^{-q}/\partial v) = \beta(-x^p y^{-q}(1 + y + \ldots + y^{q-1})(-l^{-1})\partial l/\partial v)$$

and

$$\beta(\partial R/\partial x) = \beta(\partial x^p (l^{-1}x^r)^{-q}/\partial x).$$

The next step is to obtain a representation ρ of $\mathbb{Z}G$ into some matrix ring over a commutative ring A with $0 \neq 1 \in A$, which maps $I_{\{x,v\}}$ to 0. The fundamental formula for the Fox calculus gives

$$0 = \beta(R) - 1 = \beta(\partial R/\partial x)(x-1) + \beta(\partial R/\partial v)(v-1).$$

Hence it is sufficient to find a representation $\rho : \mathbb{Z}G \to \mathbb{M}_m(A)$ where the image of $\beta(\partial R/\partial v)$ equals 0 and $\rho(x) - 1$ is not a zero divisor. For example, if we choose $m = 1$ and $A = \mathbb{C}$, then any homomorphism ρ of G with $\rho(y) = e^{2\pi i/q}$ and $\rho(x) \neq 1$ will define a representation of $\mathbb{Z}G/I$. Since the abelianization of the relator R gives $xl^q = 1$, we obtain such a homomorphism ρ if x is mapped to $e^{2\pi i/p}$ and v to $e^{-2\pi i/pqw}$, where w ($\neq 0$ by assumption) is the exponent sum of v in $l = l(v, x^p)$.

The last step is to compute the Jacobian matrix $\partial\{y,v\}/\partial\{x,v\}$, its image $[\partial\{y,v\}/\partial\{x,v\}]$ as a matrix over $\mathbb{C}$, and its determinant. It is immediate that

$$\frac{\partial\{y,v\}}{\partial\{x,v\}} = \begin{pmatrix} \partial y/\partial x & 0 \\ * & 1 \end{pmatrix},$$

so that $\det[\partial\{y,v\}/\partial\{x,v\}]$ is the image of $\partial y/\partial x$ in $\mathbb{C}$. We compute

$$\partial y/\partial x = \partial l^{-1}x^r/\partial x = -l^{-1}(\partial l/\partial x) + l^{-1}(1 + x + \ldots + x^{r-1}).$$

Since l is a word in x^p and v, we obtain $(\partial l/\partial x) = (\partial l/\partial x^p)(\partial x^p/\partial x) = (\partial l/\partial x^p)(1 + x + \ldots + x^{p-1}) \to 0$ as $x \to e^{2\pi i/p}$. Hence the absolute value of $\det[\partial\{y,v\}/\partial\{x,v\}]$ is $|(1 - x^r)/(1 - x)|$. This is strictly bigger than 1, since $1 < r < p - 1$, by assumption on p, q and r. As the modulus of the image of x and v is 1, the determinant $\det(\partial\{y,v\}/\partial\{x,v\})$ can not be a trivial unit. Thus by Theorem 0.1 the generating systems $\{x,v\}$ and $\{y,v\}$ are not Nielsen equivalent.

Now we claim that $\phi(x) = x$ and $\phi(v) = lvl^{-1}$ defines an automorphism $\phi : G \to G$. This can be seen either by a direct computation, or by a general argument for amalgamated products, since l is contained in the center of the amalgam (see for example [CL]). An easy computation shows that $\phi(l) = l$ and hence $\phi^n(x) = x$, $\phi^n(v) = l^n v l^{-n}$. Notice that $\phi(y) = y$ and thus $\phi^n(y) = y$.

Proposition 2.2 *For all $n \in \mathbb{Z}$ the generating system $\{y,v\}$ is Nielsen inequivalent to the generating system $\{\phi^n(x), \phi^n(v)\}$, and the generating system $\{x,v\}$ is Nielsen inequivalent to $\{\phi^n(y), \phi^n(v)\}$.*

Proof. The fundamental formula for Fox derivatives gives

$$l^n - 1 = \partial l^n/\partial v(v-1) + \partial l^n/\partial x(x-1) = \partial l^n/\partial v(v-1) + (\partial l^n/\partial x^p)\, \partial x^p/\partial x(x-1).$$

When evaluated in $\mathbb{C}$ via the map ρ, this becomes equal to $\rho(\partial l^n/\partial v(v-1))$, as $\rho(x) = e^{2\pi i/p}$. Now consider the determinant

$$\det \rho \left(\frac{\partial\{\phi^n(x), \phi^n(v)\}}{\partial\{x,v\}} \right).$$

It is immediate that this is equal to

$$\rho(\partial l^n v l^{-n}/\partial v) = \rho((1 - l^n v l^{-n})\partial l^n/\partial v + l^n) = 1.$$

Recall that the torsion invariant satisfies a cancellation rule as stated in Remark 1.9 (2). Thus Proposition 1.8 gives

$$
\begin{aligned}
&\mathcal{N}(\rho)(\mathcal{N}(\{\phi^n(x), \phi^n(v)\}, \{y, v\})) \\
={}& \mathcal{N}(\rho)(\mathcal{N}(\{\phi^n(x), \phi^n(v)\}, \{x, v\})) + \mathcal{N}(\rho)(\mathcal{N}(\{x, v\}, \{y, v\})) \\
={}& \mathcal{N}(\rho)(\mathcal{N}(\{x, v\}, \{y, v\})) \\
={}& -\mathcal{N}(\rho)(\mathcal{N}(\{y, v\}, \{x, v\})),
\end{aligned}
$$

which is non-zero by the proof of Proposition 2.1. Hence $\{\phi^n(x), \phi^n(v)\}$ and $\{y, v\}$ are not Nielsen equivalent. The second statement is proved the same way. $\quad\square$

3. Topological applications

Let $M(\alpha, \beta, p, q)$, with $\gcd(p, q) = \gcd(\alpha, \beta) = 1$, $\alpha > 4$ and even, be the three-manifold which is obtained by gluing the complement of the two-bridge link $K = K(\alpha, \beta) = K_1 \cup K_2$ along the boundary component $\partial N(K_2)$ to the complement of the torus knot $T(p, q)$ in the following way: the gluing map sends the meridian u of K_2 to the fiber of the Seifert fiberation of $S^3 - N(T(p, q))$. A longitude l of K_2 is mapped to a cross curve of the Seifert fiberation. The manifolds $M(\alpha, \beta, p, q)$ are the complements of non-simple knots with tunnel number 1, as shown by Morimoto and Sakuma (see [MSa]). The fundamental group of the knot exterior, $G = \pi_1(M(\alpha, \beta, p, q))$, has a presentation as a free product with amalgamation over a $\mathbb{Z} \oplus \mathbb{Z}$ which is of the type considered in section 2 (not neccessarily satisfying the conditions (1) and (2) above), and with the additional specification that

$$l = v^{\epsilon_1} u^{\epsilon_2} \ldots u^{\epsilon_{\alpha-2}} v^{\epsilon_{\alpha-1}}, \qquad \epsilon_i = -1^{\lfloor i\beta/\alpha \rfloor}.$$

Here the generators x, y are represented by the singular fibers in the Seifert fiberation of $S^3 - N(T(p, q))$, and v is a meridian on the boundary component $\partial N(K_1)$. Similarily as in section 2, one can see that this group is a two-generator, one-relator group. (Alternatively, this follows from [MSa]).

However, in order to apply the results from the previous section, we will concentrate on the case

$$p - rq = 1, \quad p, q, r > 1, \quad \epsilon_1 + \epsilon_3 + \ldots + \epsilon_{\alpha-1} \neq 0,$$

which gives us a presentation for G precisely as in the previous section.

By a Heegaard splitting for a three-manifold M with boundary we mean a decomposition of M into a handlebody and a union of two-handles. From such a Heegaard splitting we can deduce a presentation for the fundamental group of M. If two Heegaard splittings for M are isotopic, then the isotopy takes one handlebody to the other and thus induces a Nielsen equivalence between the generating systems of the corresponding presentations for $\pi_1(M)$.

In particular, the generating systems $\{x, v\}$, $\{y, v\}$ of $G = \pi_1(M(\alpha, \beta, p, q))$ come from two Heegaard splittings $\Sigma\{x, v\}$, $\Sigma\{y, v\}$ of $M(\alpha, \beta, p, q)$ which were exhibited in [MSa] (and denoted there by $\tau(j, z, 1)$, with $j \in \{1, 2\}$ and $z \in \{x, y\}$).

The automorphisms ϕ^n above are induced on G by an n-fold Dehn twist along the incompressible torus in the knot space. This Dehn twist takes the Heegaard splittings $\Sigma_{\{x,v\}}$, $\Sigma_{\{y,v\}}$ of $M(\alpha, \beta, p, q)$ to Heegaard splittings $\Sigma^n_{\{x,v\}}$, $\Sigma^n_{\{y,v\}}$ (which are denoted $\tau(i, z, n)$ in [MSa]). The above discussion, together with Proposition 2.2, proves:

Corollary 3.1. *For all $n, m \in \mathbb{Z}$ none of the Heegaard splittings $\Sigma^n_{\{x,v\}}$ is isotopic to any of the $\Sigma^m_{\{y,v\}}$.*

This corollary is proved by different methods in [MSa] (Theorem 4.1 (3)). In fact, their elaborate arguments give a complete isotopy and homeomorphism classification of all genus two Heegaard splittings of the manifolds $M(\alpha, \beta, p, q)$. This seems to be difficult to obtain via $\mathcal{N}(G)$ in full generality. On the other hand, the computations performed in the previous sections, based on $\mathcal{N}$-torsion, are comparatively fast and easy. Furthermore, one should notice that:

(1) The class of groups considered in Proposition 2.2 is more general than just three-manifold groups.

(2) Even for the case of three-manifold groups the algebraic result (about Nielsen equivalence), given in Proposition 2.2, does not follow from the geometric statements in Corollary 3.1 or in [MSa].

References

[BZ] G. Burde and H. Zieschang, *Knots*, Studies in Mathematics 5, de Gruyter, New York, 1985.

[CL]   M. M. Cohen and M. Lustig, *Very small group actions on ℝ-trees and Dehn twist automorphisms*, preprint.

[F]   R. H. Fox, *Free differential calculus I. Deriviation in the free group ring*, Ann. of Math. **57** (1953), pp. 547–560.

[L]   M. Lustig, *On the Rank, the Deficiency and the Homological Dimension of Groups: The Computation of a Lower Bound via Fox Ideals*, in: Topology and Combinatorial Group Theory (P. Latiolais, ed.), Lecture Notes in Mathematics 1440, Springer, 1991, pp. 164–174.

[LM1]   M. Lustig and Y. Moriah, *Nielsen Equivalence in Fuchsian Groups and Seifert Fibered Spaces*, Topology **30** (1991), pp. 191–204.

[LM2]   M. Lustig and Y. Moriah, *Generating Systems for Groups and Reidemeister-Whitehead Torsion*, to appear in J. of Algebra.

[LM3]   M. Lustig and Y. Moriah, *Generalized Montesinos Knots, Tunnels and $\mathcal{N}$-Torsion*, to appear in Math. Annalen.

[LS]   R.C. Lyndon and P.E. Schupp, *Combinatorial Group Theory*, Modern Studies in Math. 89, Springer.

[M]   J. Milnor, *Algebraic K-Theory*, Ann. of Math. Study 72, Princeton University Press.

[MKS]   W. Magnus, A. Karrass and D. Solitar, *Combinatorial Group Theory*, Interscience, New York, 1966.

[MoS]   Y. Moriah and V. Shpilrain, *Non-tame automorphisms of extensions of periodic groups*, preprint.

[MSa]   K. Morimoto and M. Sakuma, *On unknotting tunnels for knots*, Math. Ann. **289** (1991), pp. 143–167.

Surface Groups and Quasi-Convexity

Christophe Pittet

Section de Mathématiques, Université de Genève, 2-4 rue du Lièvre, C.P. 240, 1211 Genève 24, Switzerland.

Introduction

Using the notion of a quasi-convex subgroup, developed by M. Gromov [Gr], H. Short [Sh], gives a geometric proof of a theorem of Howson [Ho]: in a free group the intersection of two finitely generated subgroups is again finitely generated. We show that the same geometric approach applies to surface groups, the key point being that, in a hyperbolic surface group, a subgroup is finitely generated if and only if it is quasi-convex. Translated to the context of rational structures on groups, this fact gives a positive answer to a question of Gersten and Short [GS] about rational subgroups.

We denote by $\mathcal{C}(\Gamma, S)$ the Cayley graph of Γ relative to S where Γ is a group generated by a finite symmetric set $S = S^{-1}$. We denote by d_S the left invariant word metric on Γ relative to S (see [GH] or [CDP] for definitions).

1. Quasi-convex subgroups

Definition. (Gromov [Gr 5.3, p. 139]) *Let Γ be a group generated by a finite set $S = S^{-1}$. A subgroup H of Γ is quasi-convex with respect to S if there exists a constant K such that any geodesic segment of the Cayley graph $\mathcal{C}(\Gamma, S)$ joining two points of H stays in a K-neighbourhood of H.*

Remark 1. *If $\Gamma = F(a_1, \ldots, a_n)$ is the free group of rank n with its natural generator system $S = \{a_1^{\pm 1}, \ldots, a_n^{\pm 1}\}$, then any finitely generated subgroup H of Γ is quasi-convex with respect to S.*

Proof. $\mathcal{C}(\Gamma, S)$ is a tree. For any $h \in H$ there exists exactly one geodesic between the identity element e and h. Suppose that H is finitely generated by T, so that $h = t_1 t_2 \cdots t_k$ where $t_i \in T$. The geodesic is covered by the geodesic segments joining $h_i = t_1 \cdots t_i$ to $h_{i+1} = h_i t_{i+1}$ (say $h_0 = e$). These segments are of uniformly bounded length because T is finite, and their end points lie on H.

Remark 2.   *If Γ is finitely generated by S and if H is a quasi-convex subgroup of Γ with respect to S then H is finitely generated.*

Proof. Take $h \in H$ and choose any geodesic of $\mathcal{C}(\Gamma, S)$ joining e to h. The geodesic corresponds to a decomposition $h = s_1 \cdots s_n$, where $s_i \in S$. Let K be the quasi-convexity constant. For every point $\gamma_i = s_1 \cdots s_i$ on the geodesic there exists $h_i \in H$ such that $d_S(\gamma_i, h_i) \leq K$, and $h_n = h$. Then

$$h = h_1(h_1^{-1}h_2) \cdots (h_n^{-1}h_n),$$

and $d_S(e, h_i^{-1}h_{i+1}) = d_S(h_i, h_{i+1}) \leq 2K + 1$, thus H is generated by a set contained in the finite ball $B_{2K+1}(e)$ of $\mathcal{C}(\Gamma, S)$.

Proposition 1. (Short) *If Γ is finitely generated by S and if A and B are quasi-convex subgroups of Γ with respect to S, then $A \cap B$ is also quasi-convex with respect to S.*

The theorem of Howson follows easily from Proposition 1: if A and B are finitely generated subgroups of a free group — we can assume that the free group is finitely generated by considering the subgroup generated by $A \cup B$ — they are quasi-convex by Remark 1, so $A \cap B$ is quasi-convex, and by Remark 2, also finitely generated.

Proof of Proposition 1. Let K_A and K_B be the quasi-convexity constants for A and B. Set

$$E = \{(\gamma, \eta) \in \Gamma \times \Gamma \mid d_S(e, \gamma) \leq K_A, \ d_S(e, \eta) \leq K_B\}$$

and let N denote the number of elements of E. We will show that $A \cap B$ is N-quasi-convex with respect to S. Take $h \in A \cap B$, choose any geodesic path joining e to h and let $h = s_1 \cdots s_n$ be the corresponding decomposition. Let z be a point on this path. Suppose the path contains a segment of length N which contains h but not z; otherwise there is nothing to show. By hypothesis, starting from each point $s_1 \cdots s_k$ on the geodesic, there exists a geodesic segment described by an element $\gamma_k \in \Gamma$ (resp. $\eta_k \in \Gamma$) of length less or equal to K_A (resp. K_B) with end point in A (resp. B) and we can choose $\gamma_n = \eta_n = e$. There are $N + 1$ pairs $(\gamma_k, \eta_k) \in E$ for $n - N \leq k \leq n$, thus there must exist two *distinct* numbers i and j with $n - N \leq i < j \leq n$ such that $(\gamma_i, \eta_i) = (\gamma_j, \eta_j)$. This implies that

$$h' = s_1 \cdots s_i s_{j+1} \cdots s_n = s_1 \cdots s_i \gamma_i \gamma_j^{-1} s_{j+1} \cdots s_n = s_1 \cdots s_i \eta_i \eta_j^{-1} s_{j+1} \cdots s_n$$

is an element of $A \cap B$.

Note that this *shorter* path from e to h' described by $s_1 \cdots s_i s_{j+1} \cdots s_n$ is not in general a geodesic. But, starting from each of its vertices there still exists a path, equal to a γ_k (resp. η_k) or a translated version of a γ_k (resp. η_k), of length less than or equal to K_A (resp. K_B) with end point in A (resp. B). This means that if the segment joining z to h' on this shorter path is still longer than or equal to N we can apply the same trick again to obtain $h'' \in A \cap B$ and so on. We will eventually come up with $h^{(m)} \in A \cap B$ at a distance less than or equal to N from z.

2. Surface groups

Proposition 2. *Let Γ be the fundamental group of a closed surface X which admits a hyperbolic structure. If H is a finitely generated subgroup of Γ then H is quasi-convex in Γ (with respect to any finite generator system of Γ).*

This implies:

Proposition 3. *Let Γ be the fundamental group of a closed surface. If A and B are finitely generated subgroups of Γ then $A \cap B$ is again finitely generated.*

Proof of Proposition 2. We denote by $\mathbb{H}$ the two dimensional hyperbolic space. Let $Y = H \backslash \mathbb{H}$ be the covering of X corresponding to H. Topologically, Y is a closed surface minus a finite (possibly empty) collection of disjoint closed disks D_i. Furthermore, Y is negatively curved and without cusps; there are no non-trivial small loops in Y because it covers a compact space. Let γ_i be a loop in Y going around D_i, then for each i there exists a unique geodesic loop in the free homotopy class of γ_i. These geodesic loops are disjoint because the D_i are disjoint. As a consequence, there exists a *compact geodesically convex* submanifold $C \subseteq Y$ on which Y retracts by deformation. The covering projection

$$p : \mathbb{H} \to Y$$

is a local isometry so that $Z = p^{-1}(C)$ is geodesically convex in $\mathbb{H}$. Pick a base point x_0 in $\mathbb{H}$ and let

$$\varphi : \Gamma \to \mathbb{H}, \qquad \gamma \mapsto \gamma x_0,$$

be the monodromy action. This induces a commutative diagram:

where the vertical arrows are inclusion maps. The group Γ (resp. H) acts properly and discontinuously by isometries on $\mathbb{H}$ (resp. Z) with compact quotient X (resp. C). It follows that the horizontal arrows are quasi-isometry maps (see [GH III.3.19]). Because Z is convex, H is quasi-convex with respect to any finite generator system of Γ (see [CDP 10.1.4]).

Remarks. (i) Z is the convex hull of the limit set of H.
(ii)   It is a well-known fact that in a hyperbolic group a subgroup which is quasi-convex with respect to a particular finite generator system is also quasi-convex with respect to any finite generator system (see [CDP 10.4.1]).

3. An example in dimension three

Let us recall classical examples of Jørgensen and Thurston which show that there exist finitely generated subgroups in hyperbolic three-manifold groups which are not quasi-convex. Let M be a closed hyperbolic three-manifold which fibres over the circle with fibre a closed surface of genus two (see [Su]). Let Γ be the fundamental group of M and let S be a finite generator system for Γ. Let H be the fundamental group of the fibre and let T be a finite generator system for H. We assume that $T \subset S$. Let $\mathbb{H}$ now be the three dimensional hyperbolic space. Let us pick a base point $x_0 \in \mathbb{H}$. As above the monodromy action gives a quasi-isometry $\mathcal{C}(\Gamma, S) \to \mathbb{H}$ and we obtain a commutative diagram:

$$
\begin{array}{ccc}
\mathcal{C}(\Gamma, S) & \longrightarrow & \mathbb{H} \\
\uparrow & & \uparrow \\
\mathcal{C}(H, T) & \longrightarrow & \mathbb{H}
\end{array}
$$

where the left vertical arrow is the inclusion map and the right vertical one is the identity map.

Now suppose H is quasi-convex in Γ. We will obtain a contradiction. At infinity we obtain a topological commutative diagram:

$$
\begin{array}{ccc}
\partial\Gamma & \longrightarrow & L_\Gamma \\
\uparrow & & \uparrow \\
\partial H & \longrightarrow & L_H
\end{array}
$$

where ∂H and $\partial \Gamma$ are Gromov boundaries (see [CDP 10.4.1] or [GH]), and where L_H and L_Γ are limit sets in the two-sphere. The vertical arrows are inclusion maps and the horizontal ones are homeomorphisms. As H is normal, L_H is Γ-invariant, but the action of Γ on the sphere is minimal (see [Th 8.1.2]), so $L_H = L_\Gamma$. This implies that

$$
S^1 \simeq \partial H \simeq L_H = L_\Gamma = S^2,
$$

which is absurd.

4. Remarks

(1)   In the case of hyperbolic surface groups, the same argument as in §3 above, combined with Proposition 2 and the fact that the boundary of a free group is a Cantor set, shows that the only non-trivial finitely generated normal subgroups are of finite index. In fact Mihalik and Lustig (see [ABC]) show that if

$$1 \to H \to \Gamma \to Q \to 1$$

is an exact sequence of infinite groups with Γ hyperbolic, then H is not quasi-convex in Γ.

(2)   There is an interesting question about a possible dichotomy between normality and quasi-convexity for finitely generated subgroups of infinite index in hyperbolic groups (see [Sw]).

(3)   Note that the finitely generated intersection property fails in the example of §3 because there exists a subgroup B generated by two elements such that $B \cap H$ is free of infinite rank (see [Ja 5.19.d]).

(4)   Let us call a hyperbolic group *spiky* if all its finitely generated subgroups are quasi-convex. It is easy to show that if A is a hyperbolic spiky group and if B is a group which is commensurable with A, then B is also hyperbolic and spiky. Using covering arguments it is possible to show that if A and B are hyperbolic spiky groups then the free product $A * B$ is also hyperbolic and spiky.

5. Rational structures on groups

We now recall some definitions and properties of rational structures on groups (see [GS]).

Definition.   *A rational structure $(\mathcal{A}, L)$ for a group G comprises*

 (i) *a finite set $\mathcal{A}$ and a map $\mu : \mathcal{A} \to G$ such that the monoid homomorphism induced by μ from the free monoid $\mathcal{A}^*$ to G is surjective;*

 (ii) *a sublanguage $L \subset \mathcal{A}^*$ which is regular (i.e. which is the accepted language of a finite state automaton), such that μ restricted to L is still surjective.*

Example.   According to Cannon and Gromov, (see [GH]), if G is word hyperbolic and if $\mathcal{A}$ is any finite set of semi-group generators, there exists a sublanguage L of the geodesic language which is regular and such that μ restricted to L is bijective.

Definition.   *A subset $B \subset G$ is L-rational if there exists a rational structure $(\mathcal{A}, L)$ for G such that $\mu^{-1}(B) \cap L$ is regular in $\mathcal{A}^*$.*

Given a rational structure $(\mathcal{A}, L)$ for G, a subset B is L-quasi-convex if there exists a constant K such that if $w \in L \cap \mu^{-1}(B)$, the path in $\mathcal{C}(G, \mathcal{A})$ corresponding to w lies in a K-neighbourhood of B.

The fundamental example is when G is word hyperbolic, words in the language are geodesics and B is a quasi-convex subgroup of G.

Theorem. (Gersten and Short [GS]) *Let $(\mathcal{A}, L)$ be a rational structure for G, let H be a subgroup of G. Then H is L-rational if and only if H is L-quasi-convex.*

Gersten and Short [GS] have asked if there exists a rational structure on the fundamental group of a negatively curved Riemannian surface such that every finitely generated subgroup is rational. According to Proposition 2, to the result of Cannon and Gromov and to the theorem above, we can see that this is the case. More generally, (see [Sw]):

Theorem. (Swarup) *Let G be a geometrically finite torsion-free Kleinian group with non zero Euler characteristic and without parabolics. Let $(\mathcal{A}, L)$ be any rational structure on G which is biautomatic. A subgroup of G is L-rational if and only if it is finitely generated.*

Acknowledgements. I am very grateful to H. Short and G.A. Swarup from whom I have learnt these ideas about quasi-convexity, to E. Artal and P. de la Harpe for help and criticism and also to M.A. Roller and G.A. Niblo who organised the 1991 Geometric Group Theory meeting at Sussex.

References

[ABC] J.M. Alonso, T. Brady, D. Cooper, V. Ferlini, M. Lustig, M. Mihalik, M. Shapiro and H. Short, *Notes on word hyperbolic groups*, in: Group Theory from a Geometrical Viewpoint (A.Haefliger, E. Ghys and A. Verjovsky, eds.), World Scientific, 1991.

[CDP] M. Coornaert, T. Delzant and A. Papadopoulos, *Géometrie et théorie des groupes*, Lectures Notes in Math. 1441, Springer, 1990.

[GH] E. Ghys and P. Harpe, *Sur les groupes hyperboliques d'après M. Gromov*, Birkhäuser, 1990.

[Gr] M. Gromov, *Hyperbolic groups*, in: Essays in Group Theory (S.M. Gersten, ed.), MSRI series 8, Springer, 1987 pp. 75–264.

[GS] S.M. Gersten and H. Short, *Rational subgroups of biautomatic groups*, Annals of Math. **134** (1991), pp. 125–158.

[Ho] A.G. Howson, *On the intersection of finitely generated free groups*, J. London Math. Soc. **129** (1954), pp. 428–434.

[Ja] W. Jaco, *Lectures on three-manifold topology*, A.M.S. Reg. Conf. Series in Math. 43, (1980).

[Sh]   H. Short, *Quasi-convexity and a theorem of Howson*, in: Group Theory from a Geometrical Viewpoint (A.Haefliger, E. Ghys and A. Verjovsky, eds.), World Scientific, 1991.

[Su]   D. Sullivan, *Travaux de Thurston sur les groupes quasi-fuchsiens et les variétés hyperboliques de dimension 3 fibrées sur S^1*, in: Séminaire Bourbaki, exposé 554, 32e année 1979/80, Lectures Notes in Math. 842, Springer, 1980 pp. 196–214.

[Sw]   G.A. Swarup, *Geometric finiteness and rationality*, research report, University of Melbourne, 1990.

[Th]   W. Thurston, *The Topology and Geometry of Three-manifolds*, preprint, Princeton, 1980.

Constructing Group Actions on Trees

Martin A. Roller

Mathematik, Universität Regensburg, Postfach 397, 8400 Regensburg, Germany.

Introduction

The theory of Bass and Serre [Se] of groups acting on trees has turned out to be an extremely useful tool in Combinatorial Group Theory. It relates two of the most basic, but unfortunately rather inconsistently named constructions: amalgamated products and HNN-extensions, and makes their algebraic and combinatorial properties completely transparent by means of an associated group action on a tree. In fact, Bass and Serre show that the general theory of groups acting on trees is the same as the theory of groups obtained by repeated applications of those two basic constructions.

In order to employ this powerful theory one needs a plentiful supply of groups that admit interesting actions on trees. Typical examples studied in Serre's book are the linear groups of dimension two over $\mathbb{Z}$ or the field of fractions of a principal valuation ring.

One of the oldest methods to recognize free products [LS, Gl] is known in the mathematical lore as

Ping Pong Lemma. *Let G be a group with two subgroups G_1 and G_2, let M be a G-set containing two disjoint nonempty subsets M_1 and M_2 such that*

(i) $gM_1 \subseteq M_2$ *for all* $g \in G_1 - \{1\}$,
(ii) $gM_2 \subseteq M_1$ *for all* $g \in G_2 - \{1\}$.

Then the group generated by G_1 and G_2 is either the free product $G_1 * G_2$ or a dihedral group.

The second case only arises when both G_1 and G_2 are cyclic of order two and equality holds in (i) and (ii). Suppose G_1 contains two distinct nontrivial elements g and h, then $g^{-1}hM_1 \subseteq M_2$, hence gM_1 and hM_1 are disjoint proper subsets of M_2. Inductively, one can show that the family

$$(M_1, M_2, g_ng_{n-1}\cdots g_1M_1, g_ng_{n-1}\cdots g_2M_2 \mid g_{2i+1} \in G_1 - \{1\}, g_{2i} \in G_2 - \{1\})$$

consists of distinct subsets of M, which immediately implies that G_1 and G_2 generate their free product.

In fact, the proof shows a little more: any two members of that family either are disjoint, or one contains the other. If one pictures the usual planar Venn diagram of this family, the Bass-Serre tree appears visible to the naked eye as the dual graph of this diagram.

Stallings [St] reformulated and generalized this method in terms of bipolar structures and used the action of the group on itself and the shape of its Cayley graph to give an algebraic characterization of finitely generated groups which act on a tree with finite edge stabilizers. The connection between Stalling's proof and Bass-Serre Theory was later made explicit by Dunwoody [Du]. The monograph by Dicks and Dunwoody [DD] contains powerful generalizations of this result and a wealth of applications.

A generalization in a different direction, which puts stronger restrictions on the group but allows the edge stabilizers to vary in a class of infinite groups, forms the basis of a series of papers by Kropholler and Roller [KR1-4], which culminates in Kropholler's proof the the Torus Theorem [K1, K2].

In this note we survey the construction of a tree with a group action used in all the generalizations of Stallings' work ([Du], [DD], [KR], [DR]). But first we start of with a classification of group actions on trees that goes back to Tits [Ti].

Acknowledgements. This paper grew out of numerous conversations with Peter Kropholler and Martin Dunwoody. I would like to thank them both for sharing their insights with me.

1. Types of group actions on trees

1.1. A *graph X* consists of two disjoint sets, the *vertices VX* and *edges EX*, together with incidence maps $\iota, \tau : EX \to VX$, denoting the *initial* and *terminal* vertex of an edge and a fixed point free involution $* : E \to E$, such that $\iota e^* = \tau e$ and $\tau e^* = \iota e$ for all $e \in EX$. The pair $\{e, e^*\}$ represents the two different orientations of the (unoriented) edge joining the vertices ιe and τe. An *orientation* of T is a subset E^+ of EX such that for every $e \in EX$ exactly one of e and e^* lies in E^+.

If a group G acts on VX and EX such that ι, τ and $*$ are G-maps, then X is called a *G-graph*. We say that G *preserves orientations* if there exists an orientation E^+ which is a G-set, otherwise G *acts with involutions*, i.e. there exists an $e \in EX$ and a $g \in G$ with $ge = e^*$.

A *path* of length n in X is a sequence $e_1, \ldots, e_n \in EX$ with $\tau e_i = \iota e_{i+1}$, and it is *reduced* if $e_i^* \neq e_{i+1}$, for $i = 1, \ldots, n-1$. The graph X is *connected* if any two vertices can be connected by a path, and the *distance* of these vertices is the length of the shortest path connecting them.

1.2. A *tree* T is a connected graph that does not contain any cycles, i.e. any reduced path beginning and ending with the same vertex. This means that for any two vertices in VT there exists a unique reduced path, or *geodesic*, connecting them. There are many more ways of characterizing trees (see e.g. [DD] I.6). For our purposes the most useful property is:

> A connected graph T is a tree, if for any $e \in ET$ the graph $T-\{e, e^*\}$ consists of precisely two connected components.

We denote the component containing ιe by T_e^ι, and the other component by T_e^τ.

Later we shall use two further relations to describe the structure of a tree (cf. [Du] and [Di]). For an edge e and a vertex v we say that e *points to* v, and write $e \to v$, if $v \in T_e^\tau$. Otherwise, if $v \in T_e^\iota$, we say that e *points away from* v and write $v \to e$.

For edges $e, f \in ET$ we define the ordering relation $e \geq f$ (*e before f*) if $T_e^\tau \supseteq T_f^\tau$, i.e. if there exists a geodesic in T beginning with e and ending with f. It is clear that if T is a G-tree then both these relations are G-equivariant. The key observation of Dunwoody [Du] is that the following properties of the partially ordered set ET also characterize trees.

(i) For any $e, f \in ET$ precisely one of the following holds

$$e = f, \quad e = f^*, \quad e > f, \quad e^* > f, \quad e > f^*, \quad e^* > f^*.$$

(ii) For any $e, f \in ET$ the set $\{h \in ET \mid e < h < f\}$ is finite.

A partially ordered set that admits an order reversing involution * and satifies just (i) is said to be *nested*, while (ii) is called the *finite interval condition*.

1.3. Now we consider a tree T with a G-action.

Lemma and Definition. *We say that an edge $e \in ET$ splits T if one of the following equivalent conditions holds.*

(i) For some $v \in VT$ both sets $T_e^\iota \cap Gv$ and $T_e^\tau \cap Gv$ have infinite diameter.

(ii) For every $v \in VT$ both sets $T_e^\iota \cap Gv$ and $T_e^\tau \cap Gv$ have infinite diameter.

(iii) Both T_e^ι and T_e^τ contain translates of e.

(iv) There is a geodesic in T which contains three different translates of e.

(v) Some element $g \in G$ shifts e, i.e. either $e < ge$ or $e > ge$.

Proof. Let $v_1, v_2 \in VT$ be two vertices at distance d. Then any vertex in the orbit Gv_1 is at distance d from a vertex in Gv_2, so (i) implies (ii).

If (ii) holds, then both T_e^ι and T_e^τ contain translates of ιe which have distance at least 2 from ιe. Therefore both T_e^ι and T_e^τ contain translates of e.

Any geodesic through translates of e in both T_e^ι and T_e^τ must also contain e, so (iv) holds.

Given three translates g_1e, g_2e, g_3e on a geodesic, then at least two of them must be comparable, say $g_1e < g_2e$. This means that $g_1^{-1}g_2$ shifts e. Finally, if g shifts e, then all translates $\{g^ne \mid n \in \mathbb{Z}\}$ are distinct and lie on a bi-infinite geodesic, in particular (i) is satisfied for $v = \iota e$. □

1.4. We distinguish three kinds of G-trees.

a) T is called *elliptic* if for some $v \in VT$ the orbit Gv has finite diameter.

b) T is called *hyperbolic* if some edge splits it.

c) T is called *parabolic* if it is neither elliptic nor hyperbolic.

Obviously, if some vertex orbit Gv has finite diameter, then the same is true for all other vertex orbits. It is well known [DD, I.4.9] that a G-tree is elliptic if and only if there is either an edge pair $\{e, e^*\}$ or a vertex fixed under the action of G.

Now suppose T is a hyperbolic tree. Let $\hat{T}$ denote the subgraph of T comprising all edges that split T, together with all vertices incident with these edges. This is obviously a G-subgraph, and it is connected, because if e and f are splitting edges, then any edge of the geodesic connecting e to f is also splitting. Furthermore, if e is a splitting edge, then any G-subtree of T must neccessarily contain vertices in T_e^ι and T_e^τ and thus contain e. Hence $\hat{T}$ is the unique minimal G-subtree of T.

Next consider a parabolic tree. Here for every edge e and vertex v precisely one of the sets $T_e^\iota \cap Gv$ and $T_e^\tau \cap Gv$ has finite diameter. The set $E^+ = \{e \in ET \mid T_e^\iota \cap Gv$ has finite diameter$\}$ is independent of v, and it is a G-invariant orientation of T.

With this orientation it is impossible to have $e^* \geq f$ for any $e, f \in E^+$, otherwise both e and f would split T; in other words, no two edges of E^+ point away from each other. This means that for each vertex v there exists a unique edge $e_v \in E^+$ with $\iota e_v = v$ and thus a unique semi-infinite geodesic γ_v starting at v. Furthermore, for any other vertex w the geodesics γ_v and γ_w must agree on all but finitely many edges, thus the family of these geodesics defines an end of T which is fixed under the action of G. For any $e \in E^+$ the stabilizer $G_{\iota e}$ is contained in $G_{\tau e}$. The stabilizers of the vertices of a geodesic γ_v form an ascending chain whose union is G. In particular, G can not be finitely generated.

In the parabolic case there are many G-subtrees, e.g. for any $e \in E^+$ the subgraph $T_{Ge}^\tau := \bigcap_{g \in G} T_{ge}^\tau$ is a G-tree. However, there is no minimal G-subtree, because

$$\bigcap_{e \in E^+} T_{G_e}^\tau = \varnothing.$$

We summarize our observations in the following

Corollary and Definition. *If T is an elliptic G-tree, then there exists an edge or a vertex fixed under G. If T is hyperbolic, then there is a unique minimal G-subtree, and if T is parabolic, then there is no minimal G-subtree. Suppose G is generated by a single element g, then T is either elliptic or hyperbolic as a G-tree, and we call g elliptic or hyperbolic, accordingly. If g is hyperbolic, then the unique minimal g-invariant subtree of T is a bi-infinite path, called the axis of g.*

1.5. How can one tell from the structure of a group, what kind of actions on a tree it admits?

Trivially, every group admits an elliptic action on any tree. However, there are also groups which only admit elliptic actions (this is called property (FA) in Serre's book [Se]). The most obvious example is the class of finite groups. For an infinite group with that property consider $G^{p,q,r} = \langle a, b, c \mid a^p = b^q = c^r = abc = 1 \rangle$. If $G^{p,q,r}$ acts on a tree T without involutions, then there are subtrees T_a and T_b fixed pointwise under the action of a and b. A simple argument shows that if T_a and T_b are disjoint then ab is hyperbolic, contradicting the fact that c has order at most r. Thus the tree $T_a \cap T_b$ is fixed pointwise under the action of a and b, and hence all of G. It is well known that $G^{p,q,r}$ is infinite for $\frac{1}{p} + \frac{1}{q} + \frac{1}{r} \leq 1$.

Now consider the group $K^{p,q,r} = \langle a, b, c \mid a^p = b^q = c^r = abc \rangle$. This is a central extension of $G^{p,q,r}$, and it is torsion free if the latter group is infinite. Suppose that $z = abc$ acts hyperbolically on a tree T. Then the actions of a, b and c are all powers of the same automorphisms of T, which can only happen when $\frac{1}{p} + \frac{1}{q} + \frac{1}{r} = 1$. If this is not the case, then a, b, c and z all must act elliptically, and the argument above shows that there exists a vertex fixed under $K^{p,q,r}$.

Parabolic actions are easily classified. A group G admits a parabolic action on a tree if and only if it is the union of a countable ascending chain of subgroups $G_1 < G_2 < G_3 < \cdots < G$.

The most interesting case is that of hyperbolic actions. We say that G *splits* (*over a subgroup S*) if it admits a hyperbolic action on a tree (and S is the stabilizer of a splitting edge e). In this case Bass-Serre theory gives an algebraic splitting of G as an amalgamated product $G_1 *_S G_2$ or an HNN-extension $G_1 *_S$, where G_1 is the stabilizer of ιe. In the next section we shall investigate necessary and sufficient conditions on G to admit a hyperbolic action on a tree.

1.6. We finish this section with a simple example. Let $R := \mathbb{Z}[\frac{1}{2}]$ be the ring of integers extended by $\frac{1}{2}$, and let A be the group of affine self maps of R with elements $(n, r) : x \mapsto 2^n x + r$ for $r \in R$ and $n \in \mathbb{Z}$. Thus A is the semidirect product of the additive group R^+ and the infinite cyclic group generated by the multiplication by 2. Let $Z \subset R^+$ be the cyclic group generated by $1 \in R$. We define a graph T with vertex set $V = \{aZ \mid a \in A\}$, and for every vertex $v = (n,r)Z$ we define an edge e_v with $\iota e_v = (n,r)Z$ and $\tau e_v = (n-1,r)Z$.

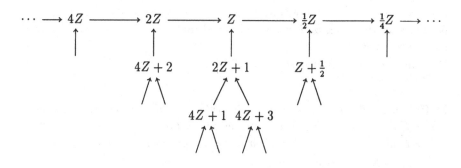

A Tree

It is easy to see that T is a tree, and each vertex is incident with three edges. The group A acts on T by left multiplication. An element $a = (n,r)$ is elliptic precisely if $n = 0$, i.e. $a \in R^+$. T is parabolic as an R^+-tree, the unique one way infinite path starting at Z is the horizontal path in the above picture, and the quotient of T under R^+ is a bi-infinite path. Viewed as an A-tree, T is hyperbolic with one orbit of vertices and edges.

2. Trees and nested sets

2.1. We are mainly interested in characterizing the subgroups over which a group may split, thus we shall restrict attention to *simple* trees, i.e. trees which have one orbit of edges under the action of G and *. Observe that a simple tree has at most two orbits of vertices. Given any G-tree T and an orbit of edges E' we can construct a simple G-tree T' with $ET' = E'$, by contracting all edges not in E'. More explicitly, for any vertex $v \in VT$ let v' denote the connected component of $T - E'$ containing v, and for $e \in E'$ define the incidence maps by $\iota'e = (\iota e)'$ and $\tau'e = (\tau e)'$.
The following is immediate from the discussion in § 1.

Proposition. *A simple G-tree T is either hyperbolic and has infinite diameter, or it is elliptic and has diameter at most 2.*

2.2. The presence of involutions in a G-action on a tree causes a slight nuisance, e.g. the exceptional case in the Ping Pong Lemma is due to this problem. To avoid it one could pass to the subgroup $G_0 \le G$ consisting of elements $g \in G$ such that for some $v \in VT$ the distance from v to gv is an even number. This subgroup has index at most two in G, and it does not contain any elements that reverse an edge of T.

Alternatively, one can form a new G-tree T' by barycentric subdivision. The vertex set of T' consists of VT plus a new vertex $v_e = v_{e^*}$ for any edge pair $e, e^* \in ET$, and there are edges e_1, e_2 with $\iota e_1 = \iota e$, $\tau e_1 = \tau e_2 = v_e$ and $\iota e_2 = \tau e$. So, if $ge = e^*$ holds in T, then g interchanges e_1 and e_2 in T'. By doubling all the edges of T we have insured that every vertex of T' is moved by an even distance.

However, in our later discussion we will have to treat this kind of tree separately.

Definition. *Let T be a simple G-tree with two orbits of vertices. A vertex v is called* special *if for every vertex w not in the orbit of v there exists an $e \in ET$ and a $g \in G$ such that $w = \tau e = \tau g e$, and w is not the terminal vertex of any other edges.*

In this situation one can always reverse the subdivision to obtain a tree $\tilde{T}$ with a single vertex orbit. In general, the two processes of subdividing and reversing are reciprocal, except in the case when T is a bi-infinite line and G acts as an infinite dihedral group. This means that G contains two conjugacy classes of inversions, one of which reverses edges and the other fixes vertices. In this case T' is also a bi-infinite line, and all vertices are special. Reversing the subdivision either yields T with the original action of G, or one where the roles of the two types of involutions are interchanged.

2.3. We now define a type of subset of G that codes the structure of T.

Definition. *For $e \in ET$ and $v \in VT$ let*

$$G[e, v] := \{g \in G \mid e \to gv\}.$$

The set $G[e, \iota e]$ is called a characteristic set *of T.*

The following proposition indicates how properties of T are reflected in properties of the $G[e, v]$'s. For subsets A and B of G we shall use $A + B$ to denote the symmetric difference, and write B^* for the complement $G - B$.

Proposition. *Let T be a simple G-tree, then for all $g \in G$, $v \in VT$ and $e \in ET$ we have*
  (i) $G[e^*, v] = G[e, v]^*$;
  (ii) $gG[e, v] = G[ge, v]$;
  (iii) $G[e, v]g = G[e, g^{-1}v]$;
  (iv) $G[e, v_1] + G[e, v_2] = \{g \in G \mid v_1 \to g^{-1}e \to v_2$ or $v_1 \to g^{-1}e^* \to v_2\}$.
  (v) T is elliptic if and only if either $G[e, v]$ or $G[e, v]^*$ is a finite union of right G_e-cosets.

(vi) Suppose that T is is hyperbolic and v is not special. Then for $e, f \in ET$
$e > f$ if and only if $G[e, v] \supset G[e, f]$.

Proof. (i)–(iv) have easy verifications.

(v) In view of (iv) the condition that either $G[e, v]$ or $G[e, v]^*$ is a finite union of right G_e-cosets for any v is equivalent to the fact that T has finite diameter, which means that it is elliptic.

(vi) Let $e, f \in ET$ and suppose that $e > f$, i.e. $T_e^\tau \supseteq T_f^\tau$, then clearly $G[e, v] \supseteq G[f, v]$. Furthermore, we have $G[e, v] \supsetneq G[f, v]$, unless $T_e^\tau \cap T_f^\iota$ consists of a single vertex $\tau e = \iota f$, which is not in the orbit of v; this means that v is special.

In this case we have $G[e, v] = G[f, v]$, there is some $g \in G$ with $e = gf^*$ and both ιe and τf are in the orbit of v. Therefore

$$G[h, v] \supset G[e, v] \quad \text{for all } h > e, \text{ and}$$
$$G[h, v] \subset G[e, v] \quad \text{for all } h < f.$$

This means that the inclusion ordering of the set $\{G[e, v] \mid e \in ET\}$ corresponds to the order on $E\widetilde{T}$. □

Remark. If G fixes v, then $G[e, v]$ is either G or empty, so we can't expect to get a nested family in this case.

2.4. Let us rephrase this in the language of [KR1]. Let S be a subgroup of G, then a subset $B \subset G$ is called *S-finite*, if $B \subseteq SF$ for some finite set $F \subset G$; B is called *S-almost invariant* if $B + Bg$ is S-finite for all $g \in G$, and it is called *proper* if neither B nor B^* is S-finite.

So, in the situation above, G acts on a simple tree T, v is a vertex of T and e is an edge with stabilizer $G_e := S$, then the set $B = G[e, v]$ has the following properties:

(i) The set of right translates of B and the orbit of v are isomorphic as G-sets, in particular $G_v = \{g \in G \mid Bg = g\}$ and B is S almost invariant.

(ii) T is a hyperbolic G-tree if and only if B is proper.

(iii) If B is proper, then the set of left translates of B and B^* corresponds to ET (or $E\widetilde{T}$), in particular it is nested with respect to inclusion and $S = \{g \in G \mid gB = B\}$.

2.5. Suppose we are given a nested set E, how can we reconstruct a tree $T(E)$ which has E as its edge set? The first construction of this kind was given by Dunwoody in [D1], where he defines a relation on E by

$$e \sim f \quad :\Longleftrightarrow \quad e = f, \text{ or } e < f^* \text{ and there is no } h \in E \text{ with } e < h < f^*.$$

If E is the edge set of a tree, then $e \sim f$ means that $\iota e = \iota f$. Conversely, Dunwoody shows that if E is just a nested set, then $\sim$ still defines an equivalence relation, and $VT(E)$ can be taken as the set of equivalence classes of

E, with $\iota e = [e]$ and $\tau e = [e^*]$. The finite interval condition then makes T into a tree.

In the approach of Dicks and Dunwoody [DD] the vertices are identified with certain orientations of the edge set. For $v \in V$ consider $A_v = \{e \in ET \mid e \to v\}$, the set of all edges pointing to v. Using just the ordering on E one can define $A_{\iota e} = \{f \in E \mid f > e \text{ or } f \geq e^*\}$. Again, they show that sets of this kind can serve as vertices for $T(E)$. Their actual construction, Theorem II.1.5 of [DD], is slightly more intricate, because it incorporates barycentric subdivisions to take care of involutions.

We shall use a variant of this method, which is adapted for simple nested sets, for which the group action together with $*$ is transitive. Here we fix an edge $e \in E$ and define a *characteristic set* $\hat{e}$ of E, which will play the role of $G[e, \iota e]$, by

$$\hat{e} := \{g \in G \mid e > ge \quad \text{or} \quad e \geq ge^*\}.$$

Since there may be two orbits of vertices, we need a candidate for $G[e, \tau e]$, which will be

$$\check{e} := (\widehat{e^*})^* = \{g \in G \mid e \geq ge \quad \text{or} \quad e > ge^*\}.$$

Now we define $VT(E) := \{\hat{e}g, \check{e}g \mid g \in G\}$ with the incidence maps $\iota ge := \hat{e}g^{-1}$ and $\tau ge := \check{e}g^{-1}$. In the next two paragraphs we will show that this definition of $T(E)$ yields indeed a tree.

2.6. The following proposition collects some simple properties of characteristic sets, modelled on §2.3. For elements $e_1, e_2 \in E$ we use the interval notation

$$[e_1, e_2[:= \{g \in G \mid e_1 \leq g^{-1}e < e_2 \quad \text{or} \quad e_1 \leq g^{-1}e^* < e_2\},$$

and define $]e_1, e_2]$, $[e_1, e_2]$ and $]e_1, e_2[$ similarly.

Proposition. *Let E be a simple nested G-set. Then for all $e \in E$ and $g \in G$*
  (i) $\widehat{ge} = g\hat{e}g^{-1}$;
  (ii) $e = ge$ implies $\hat{e} = g\hat{e} = \hat{e}g$;
  (iii) $\hat{e}^{-1} = \{g \in G \mid ge > e \text{ or } ge \geq e^*\}$;
  (iv) $\hat{e} \cap \hat{e}^{-1} = \{g \in G \mid e \geq ge^*\}$.
  (v) For $x, y \in G$ we have

$$\hat{e}x^{-1} + \hat{e}y^{-1} =]xe, ye] \cup]xe, ye^*[\cup [xe^*, ye] \cup [xe^*, ye^*[.$$

Proof. (i)–(iv) are immediate. Observe that when E is the edge set of a tree we have $A_{\iota e} \cap Ge = \hat{e}^{-1}e$.

(v) Suppose that $g \in \hat{e}x^{-1} \cap \hat{e}^*y^{-1}$. This means that

$$(e > gxe \quad \text{or} \quad e \geq gxe^*) \quad \text{and} \quad (e \leq gye \quad \text{or} \quad e < gye^*),$$

which is equivalent to

$$xe \ < g^{-1}e \le ye \quad \text{or} \quad xe \ < g^{-1}e < ye^* \quad \text{or}$$
$$xe^* \le g^{-1}e \le ye \quad \text{or} \quad xe^* \le g^{-1}e < ye^*.$$

The same calculation with x and y interchanged completes the proof. $\qquad\square$

2.7. Theorem. *Let E be a simple nested set, satisfying the finite interval condition, then $T(E)$ is a tree.*

Proof. Observe that the left stabilizer of e is contained in the right stabilizer of $\hat{e}$, therefore $T(E)$ is a well-defined graph. By definition every vertex is the endpoint of some edge, therefore we only need to consider the edges. Let $x, y \in G$ and suppose that $\iota xe = \iota ye$. This means that

$$]xe, ye] \cup \,]xe, ye^*[\, \cup \, [xe^*, ye] \cup [xe^*, ye^*[= \varnothing.$$

If xe is comparable with ye, then we must have $xe = ye$, otherwise the half open intervals would be nonempty. If xe is comparable with ye^*, then we must have $xe < ye^*$ and $]xe, ye^*] = \varnothing$. Thus we have shown that

$$xe \sim ye \quad \Longleftrightarrow \quad \hat{e}x^{-1} = \hat{e}y^{-1},$$

which is quite obviously an equivalence relation.

A similar argument shows that $\iota xe = \tau ye$ if and only if either $xe = ye^*$ or both $xe < ye$ and $]xe, ye[= \varnothing$.

It follows that reduced paths in T are precisely the unrefinable chains in E, i.e. finite sequences of the form

$$e_1 > e_2 > \ldots > e_n \quad \text{with} \quad]e_{i+1}, e_i[= \varnothing.$$

It is clear that T can not contain any closed paths; for if $\iota e_1 = \tau e_n$, then either $e_1 = e_n^*$ or $e_1 < e_n$. But this contradicts the nesting of E.

The finite interval condition is neccessary and sufficient for T to be connected.
$\qquad\square$

2.8. Finally, we consider the case that E is a family of S-almost invariant subsets of G. Here every element is very close to the corresponding characteristic set.

Lemma. *Let B be an S-almost invariant subset of G such that $\{gB, gB^* \mid g \in G\}$ is nested, then $B + \hat{B}$ is S-finite.*

Proof. Observe that $\varnothing \subset B \subset G$, thus we can choose elements $b_1 \in B$ and $b_2 \in B^*$, and there exist finite subsets F_i such that $B + Bb_i^{-1} \subseteq SF_i$. If

$B \supset gB$ then $gb_1 \in B$ and $b_2 \in B^*$, so $g \in Bb_1^{-1} \cap b_2 B^{*-1}$. If $B \supseteq gB^*$ then $gb_2 \in B$ and $b_2 \in B$, so $g \in Bb_2^{-1} \cap b_2 B^{-1}$. Hence

$$\hat{B} \subseteq (Bb_1^{-1} \cap b_2 B^{*-1}) \cup (Bb_2^{-1} \cap b_2 B^{-1})$$
$$\subseteq ((B \cup SF_1) \cap b_2 B^{*-1}) \cup ((B \cup SF_2) \cap b_2 B^{-1})$$
$$= B \cup S(F_1 \cup F_2).$$

Using the fact that B is nested with all its translates we can show by a similar argument that $\hat{B}^* \subseteq B^* \cup S(F_1 \cup F_2)$, which implies that $B + \hat{B} \subseteq S(F_1 \cup F_2)$.
□

In particular the finite interval condition holds in this situation, and the tree is hyperbolic if B is proper.

Corollary. *The group G splits over a subgroup S if and only if there exists a subset $B \subset G$ such that*
(i) B is proper S-almost invariant,
(ii) $S = \{g \in G \mid gB = B\}$,
(iii) the set $E = \{gB, gB^* \mid g \in G\}$ is nested.

Remark. The above corollary provides a proof Theorem 4.16 of [K2]. The hypotheses of that theorem imply that conditions (i)–(iii) hold and $B = \hat{B}$.

References

[DD] W. Dicks and M.J. Dunwoody, *Groups acting on graphs*, Cambridge University Press, 1989.

[Di] W. Dicks, *Groups, Trees and Projective Modules*, Lect. Notes in Mathematics 790, Springer, 1980.

[DR] M.J. Dunwoody and M.A. Roller, *Splitting groups over polycyclic-by-finite subgroups*, to appear in Bull. London Math. Soc.

[Du] M.J. Dunwoody, *Accessibility and groups of cohomological dimension one*, Proc. London Math. Soc. (3) **38** (1979), pp. 193–215.

[Gl] A.M.W. Glass, *The ubiquity of free groups*, Math. Intelligencer **14** (1992), pp. 54–57.

[KR1] P.H. Kropholler and M.A. Roller, *Splittings of Poincaré duality groups*, Math. Z. **197** (1988), pp. 421–438.

[KR2] P.H. Kropholler and M.A. Roller, *Splittings of Poincaré duality groups II*, J. London Math. Soc. **38** (1988), pp. 410–420.

[KR3] P.H. Kropholler and M. A.Roller, *Splittings of Poincaré duality groups III*, J. London Math. Soc. **39** (1989), pp. 271–284.

[KR4] P.H. Kropholler and M.A. Roller, *Relative ends and duality groups*, J. Pure Appl. Algebra **61** (1989), pp. 197–210.

[K1] P.H. Kropholler, *An analogue of the torus decomposition theorem for certain Poin-caré duality groups*, Proc. London Math. Soc. (3) **60** (1990), pp. 503–529.

[K2] P.H. Kropholler, *A group theoretic proof of the torus theorem*, these Proceedings.

[LS] Lyndon and P. Schupp, *Combinatorial Group Theory*, Springer, 1977.

[Se] J-P. Serre, *Trees*, Springer, 1980.

[St] J.R. Stallings, *On torsion-free groups with infinitely many ends*, Ann. Math. **88** (1968), pp. 312–334.

[Ti] J. Tits, *Sur le groupe des automorphismes d'un arbre*, in: Essays on Topology and related problems, Mémoires dédiés à Georges de Rham (A. Haefliger and R. Narasimhan, eds.), Springer, 1970, pp. 188–211.

Brick's Quasi Simple Filtrations
for Groups and 3-Manifolds

John R. Stallings

University of California, Berkeley, California, CA 94708, USA.

Abstract. Poénaru [P] and Casson have developed an idea about the metric geometry of the Cayley graph of a group having to do with certain problems in 3-manifold theory; this is described in [G] and in [GS]. Brick [B] has developed a non-metric condition related to this, which he calls "quasi simple filtration"; a space is qsf if it is, approximately, the union of an increasing sequence of compact, 1-connected spaces. Here we outline these notions and establish the theory in a polyhedral setting. This provides a purely group-theoretic notion of qsf which seems interesting in itself.

1. Polyhedral niceties

1.1. BASIC FACTS: The theory of finite polyhedra is a standard subject; one reference is [AH], Chapter 3. The constructions in Whitehead's paper [W] involve simplicial complexes and subdivisions of particular sorts, and this too can be considered a polyhedral reference. We shall outline some of the theory here.

A finite polyhedron P is a subset of some real vector space which can be triangulated by a finite simplicial complex K whose realization is $P = |K|$. It is sometimes more convenient to consider a cell-structure C by finitely many convex open cells of various dimensions; each such is the bounded intersection of a finite number of open half-spaces with an affine subspace; the cells in such a structure are disjoint, and the boundary of any cell is a finite union of other cells; such a structure has a triangulation obtained

This work is based on research partly supported by the National Science Foundation under Grant No. DMS-8905777 by a "barycentric subdivision" C'; a vertex is put in each cell, and a set of these form the vertices of a simplex exactly when their corresponding cells are totally ordered by the relation "is in the boundary of".

A polyhedral map $f : P \to Q$ is a function whose graph, as a subset of $P \times Q$, is a polyhedron; alternatively, such a map is a continuous function which is affine on each cell of some convex-cell structure on P; given convex-cell structures C on P and D on Q, we can refine the set of convex cells of the form $f(\sigma)$ for $\sigma \in C$ together with all convex cells of D, to get a convex-cell structure E on Q, and then determine a convex-cell structure F on D by looking at the intersections of the C-cells with preimages of the E-cells; then f will map each cell of F onto a cell of E; and then we can perform the barycentric subdivision, first on E and then compatibly on F, so that the map f now becomes a simplicial map $F' \to E'$. This construction can be generalized to apply to any finite set of polyhedra and polyhedral maps provided no polyhedron is the source of two maps.

We say that a polyhedral map $f : P \to Q$ is triangulated by triangulations K and L if $|K| = P$, $|L| = Q$, and the map f is determined by a simplicial map by using the barycentric coordinates as is customary.

To summarize:

1.2. Theorem. [Triangulations of maps] *If P and Q are compact polyhedra, $A \subset P$ and $B \subset Q$ are subpolyhedra, and $f : P \to Q$ is a polyhedral map, then: f can be triangulated by triangulations of P and Q, subcomplexes of which are triangulations of A and B; furthermore, $f(A)$ and $f^{-1}(B)$ are subpolyhedra of Q and P. Any diagram of finitely many polyhedral maps can be simultaneously triangulated, provided that each polyhedron in the diagram is the source of at most one map in the diagram.*

1.3. UNTRIANGULABLE DIAGRAMS: Here is an example of a polyhedral diagram which cannot be triangulated: Let A be a triangle with vertices v_0, v_1, v_2, and B be a copy of the interval $[0,1]$ and C be another copy of $[0,1]$; we define maps $\alpha : A \to B$ and $\beta : A \to C$, which are affine on all of A, and which map as follows: $\alpha(v_0) = 0$, $\alpha(v_1) = \alpha(v_2) = 1$, $\beta(v_0) = 0$, $\beta(v_1) = 0.5$, and $\beta(v_2) = 1$. This cannot be triangulated; as a picture shows, if it were triangulated, then there would be a vertex in A on the edge $[v_0, v_2]$ mapping by β to 0.5; the image of this in B, 0.5, would be a vertex of B; the preimage of this by α must intersect the edge $[v_0, v_1]$ in a vertex whose image by β is 0.25; etc. This would produce an infinite sequence of vertices in these supposedly finite simplicial complexes.

This makes evident that there are certain technical problems when making constructions with polyhedra by the use of simplicial things. In particular, we prove a pushout-approximation sort of result below (1.10) only under very special circumstances. (Note that the pushout of the example above, when construed in the topological category, is a two-point space with one closed point which is in the closure of the other point.)

1.4. LOCALLY COMPACT POLYHEDRA: In general we shall be interested in locally compact, second countable polyhedra and maps which are proper (i.e., the inverse image of every compact set is compact). In this situation, we have a space which is covered by a locally finite collection of compact polyhedra which fit together along subpolyhedra. Such an object is the union of an increasing sequence of compact polyhedra, each one contained in the interior of the next. If $p : A \to P$ is a covering projection, where P is a compact polyhedron, then A has the structure of locally compact polyhedron; also, if K is any simplicial triangulation of P, then this lifts to A providing A with a triangulation L, such that p is simplicial with respect to L and K.

1.5. Theorem. [Pullback] *Let $\phi : K \to L$ and $\psi : M \to L$ be simplicial maps. Then there is a simplicial pullback P whose geometric realization is the topological pullback of the topological realization.*

Proof. The construction of P is something like the construction of the product of simplicial complexes, which, we recall, involves ordering the vertices of the factors. Thus, we order the vertices of the target L by some total ordering $<$, and then compatibly order the vertices of K and of L so that on vertices the maps ϕ and ψ are monotone non-decreasing. With respect to these orderings, we define the product $K \times L$ to consist of the simplicial complex whose vertices are ordered pairs (u, v) with u a vertex of K and v a vertex of L; we give these pairs the lexicographic order so that $(u_1, v_1) < (u_2, v_2)$ means: $u_1 < u_2$, or else both $u_1 = u_2$ and $v_1 < v_2$; then a simplex of the product consists of a totally ordered set of vertices whose first coordinates form a simplex of K and whose second coordinates form a simplex of L. Now, the pullback simplicial complex P is the subcomplex of the product $K \times L$ consisting of those vertices (u, v) such that $\phi(u) = \psi(v)$, and of those simplexes in $K \times L$ spanned by such vertices.

It is now a matter of checking through the proof that $|K \times L|$ can be identified with $|K| \times |L|$, to see that $|P|$ can be identified with the topological pullback. $\square$

1.6. Corollary. *Let $\phi : K \to L$ and $\psi : M \to L$ be polyhedral maps, where K is compact and ψ is proper. Then there is a compact polyhedron P and maps $\phi' : P \to M$, $\psi' : P \to K$ which make a pullback diagram in the topological category and in the polyhedral category.*

Proof. We can restrict outselves to the case that $L = \phi(K)$ and $M = \psi^{-1}(L)$, so that the whole picture consists of compact polyhedra. We first triangulate the two maps ϕ and ψ; this will involve two simplicial complexes whose geometric realization can be identified with L; these have a common subdivision, and the original triangulations of K and M can be subdivided so that the maps ϕ and ψ are simplicial with respect to these; the point of this is to keep the triangulation of the target L the same for both maps ϕ and ψ. Then 1.5 applies to this situation. □

1.7. k-CONNECTED MAPS: A map $f : A \to B$ (of topological spaces) is said to be k-connected, if, for all possible basepoints, $f_* : \pi_\ell(A) \to \pi_\ell(B)$ is an isomorphism for $\ell < k$ and a surjection for $\ell = k$. The mapping cylinder $M(f)$ is the space obtained from the disjoint open union $(A \times [0,1]) \sqcup B$ by identifying $(a,1)$ to $f(a)$ for all $a \in A$. Thus, $M(f)$ deformation-retracts to B, and so has the same homotopy type as B, and A is included as $A \times 0$ in $M(f)$. The condition that f be k-connected is the same as to say that the pair $(M(f), A)$ is k-connected, in that the relative homotopy groups (or sets, in low dimensions) $\pi_i(M(f), A) = 0$ for dimensions $i \leq k$. Also, recall that a single space X is said to be k-connected when $\pi_i(X) = 0$ for all $i \leq k$.

1.8. Lemma. [Homotopy triads] *Let A and B be topological spaces. Suppose that $A = X_1 \cup X_2$, and $B = U_1 \cup U_2$, where U_1, U_2 are open in B. Suppose that $f : A \to B$ is continuous and that $f(X_i) \subset U_i$ for $i = 1, 2$. Let k be given; and suppose that the restrictions of f, giving maps $X_i \to U_i$ (for $i = 1, 2$) and $X_1 \cap X_2 \to U_1 \cap U_2$, are k-connected. Then the map f itself $A \to B$ is k-connected.*

Proof. By dealing with mapping cylinders, we can suppose that f embeds A as a subspace of B. Then the hypotheses are that the pairs (U_i, X_i) and $(U_1 \cap U_2, X_1 \cap X_2)$ are k-connected, and the problem is to show that the pair (B, A) is itself k-connected. One might now say, "All the obstructions are zero. QED", but here is a sketch of the details:

Consider a map $g : \Delta \to B$, where Δ is an ℓ-cell, for $\ell \leq k$, such that $g(\partial \Delta) \subset A$. Triangulate Δ so finely (by the Lebesgue covering lemma) that every simplex maps by g either into U_1 or into U_2. We call a simplex of this triangulation of Δ a U_1-simplex if g maps it into U_1, a U_2-simplex if g maps it into U_2, and a U_{12}-simplex if both are true. We homotop the map g skeleton by skeleton, and simplex by simplex, so that during the homotopy the simplexes of Δ retain their characters of being $U_{1,2,}$or $_{12}$-simplexes; and so that in the end result, each U_i-simplex gets mapped into X_i and each U_{12}-simplex gets mapped into $X_1 \cap X_2$. For instance, suppose we are at a stage g' in homotoping the map so that we are bit by bit getting the p-skeleton to be mapped into A; there is a p-simplex σ, whose boundary is mapped, by g', into A; supposing, say, that σ is a U_{12}-simplex, then its boundary consists of U_{12}-simplexes, which have been already doctored to map into $X_1 \cap X_2$. Thus, g' on σ represents some element of $\pi_p(U_1 \cap U_2, X_1 \cap X_2)$, which is supposed to be zero; we can then define a homotopy of g' on σ relative to $\partial \sigma$ so that the end result maps σ into $X_1 \cap X_2$, and carefully extend this homotopy to preserve the $U_{1,2,\text{or }12}$ character of the simplexes. In the end, we will have shown that the element of $\pi_\ell(B, A)$ represented by the map g is zero.   □

The above topological fact has some polyhedral consequences.

1.9. Theorem.   *Let $k \geq 0$. Let $f : A \to B$ be a polyhedral proper map, such that for each $x \in B$, the set $f^{-1}(x)$ is k-connected. Then the map f itself is $(k+1)$-connected.*

Proof. Since homotopy groups behave well under direct limit, we can suppose that B and hence also A are compact polyhedra. Triangulate the map f. We now build up B a simplex at a time, so that the simplexes are added in non-decreasing dimension. We prove the theorem by induction on the number of simplexes of B. The induction step goes thus: Suppose that B is obtained from B^- by adding one simplex σ whose boundary is in B^-. Define $A^- = f^{-1}(B^-)$. We note that over the interior σ°, barycentric coordinates give a structure to $f^{-1}(\sigma^\circ)$ of a product, $f^{-1}(\sigma^\circ) \approx \sigma^\circ \times f^{-1}(x)$ for any $x \in \sigma^\circ$, and the map f can be taken to be the projection onto the first factor. We define U_1 to be all of B except for one point x in σ°, U_2 to be σ°; and X_i to be $f^{-1}(U_i)$. Because of the product structure, the maps $X_2 \to U_2$ and $(X_1 \cap X_2) \to (U_1 \cap U_2)$ are projections onto one factor (homotopically either a point or a sphere) with the other factor, $f^{-1}(x)$, being k-connected; an examination of the homotopy groups of the product then shows that these two maps are $(k+1)$-connected. Then, we can find deformation retractions $X_1 \to A^-$ and $U_1 \to B^-$; by the inductive hypothesis, we know that the map $A^- \to B^-$ is $(k+1)$-connected, and so we have $X_1 \to U_1$ is $(k+1)$-connected. Then 1.8 implies that $A \to B$ is $(k+1)$-connected.   □

Now we can prove a polyhedral result which would be easy if there were a pushout in the polyhedral category. Such pushouts do not exist, and therefore this requires some technical trouble. Note that we use a double barycentric subdivision; one barycentric subdivision makes the subcomplexes we are interested in into "full" subcomplexes, i.e., completely determined by their vertices; we want to identify certain vertices to get a quotient simplicial complex; in order to get something whose homotopy situation looks like that of the topological identification space, we must barycentrically subdivide once more.

1.10. Theorem. [Pushout] *Let $k \geq 0$. Suppose that*

$$
\begin{array}{ccc}
K & \xrightarrow{\ \subset\ } & A \\
\downarrow{\scriptstyle \phi|K} & & \downarrow{\scriptstyle \phi} \\
B & \xrightarrow{\ \subset\ } & X
\end{array}
$$

is a commutative diagram of polyhedra and polyhedral maps, with A and B compact, and $K = \phi^{-1}(B)$. Suppose that $\phi|K : K \to B$ is k-connected and surjective. Then there is a polyhedron P, a polyhedral inclusion $B \subset P$, a polyhedral map $f : P \to X$, a surjective polyhedral map $\alpha : A \to P$, such that there is a commutative diagram

$$
\begin{array}{ccc}
K & \overset{\subset}{\longrightarrow} & A \\
& & \\
\downarrow{\scriptstyle \phi|K} \quad P & & \downarrow{\scriptstyle \phi} \\
& & \\
B & \overset{\subset}{\longrightarrow} & X
\end{array}
$$

with $f\alpha = \phi$ (from which it follows that $f^{-1}(B) = \alpha\phi^{-1}(B) = B$); the inclusion of B into X is the composition of f with the inclusion of B into P; and the map $|\alpha| : |A| \to |P|$ is k-connected.

Proof. Since A is compact, we can restrict ourselves to the compact sub-polyhedron $\phi(A) \subset X$, and suppose that all the polyhedra in the picture are compact. What this theorem does is to describe, in the polyhedral world, a good enough substitute for what we would get in topological world by taking A and identifying the subset K to B.

THE CONSTRUCTION OF P. By 1.2, triangulate A and X so that K and B respectively are the realizations of subcomplexes and so that ϕ is simplicial with respect to these triangulations; thus there are triangulations $T(A)$ and $T(X)$ of A and X, such that the map ϕ is given by a simplicial map also called $\phi : T(A) \to T(X)$; and we have arranged it that $T(A)$ and $T(X)$ contain, respectively, subcomplexes $T(K)$ and $T(B)$, on which the restriction of ϕ yields the map $K \to B$ which we want to use for identifying some things.

Now, we barycentrically subdivide, obtaining $\phi' : T'(A) \to T'(X)$. This makes the subcomplexes covering K and B into "full subcomplexes". In other words, we can define the simplicial complex I to consist of two vertices 0 and 1 and the 1-simplex $[0, 1]$, and get simplicial maps $\epsilon_1 : T'(A) \to I$ and $\epsilon_2 : T'(X) \to I$, such that $T'(K)$ and $T'(B)$ are the inverse images, respectively, of 0. Now we barycentrically subdivide once more, getting $\phi'' : T''(A) \to T''(X)$, $\epsilon_1 : T''(A) \to I'$, and $\epsilon_2 : T''(X) \to I'$, where I' is the subdivision of I at 0.5.

In the simplicial category, we now identify the vertices of $T'''(K)$ to those of $T'''(B)$ by means of the map ϕ'' restricted to these; and we identify simplexes of $T'''(A)$ as required. The resulting simplicial complex, $T'''(A)/(T'''(K) = T'''(B))$ is to be called $T(P)$, a triangulation of a polyhedron $P = |T(P)|$. In $T(P)$ a simplex will consist of a set of vertices which are the image of some vertex set of some simplex of $T'''(A)$. The map $f : P \to X$ is the resulting geometric realization of the simplicial map $T(P) \to T'''(X)$ which exists because $T(P)$ is a pushout in the simplicial category:

$$
\begin{array}{ccc}
T''(K) & \longrightarrow & T''(A) \\
\downarrow & & \downarrow \alpha \\
T''(B) & \longrightarrow & T(P) \\
& & \searrow \\
& & T'''(X)
\end{array}
$$

The map in this diagram, from $T''(A)$ to $T(P)$, we call α. Then, α is an identification map in the simplicial category, but its geometric realization makes some identifications on points within the open star of $T''(K)$ in $T''(A)$, so that it is not, strictly, the topological identification map.

WHY α IS k-CONNECTED. Consider the map $|\epsilon_1| : A \to |I|$. This is the same as ϵ_1'. The set $|\epsilon_1|^{-1}([0, 0.5])$ is the closed star of $T''(K)$ in $T''(A)$, and it deformation retracts to K by a simple formula involving barycentric coordinates. This defines a deformation retraction also from the closed star of $T''(B)$ in $T(P)$ to B. Furthermore, the set $|\epsilon_1|^{-1}([0, 1))$ deformation retracts to $|\epsilon_1|^{-1}([0, 0.5])$; and similarly in the polyhedron P. On $|\epsilon_1|^{-1}([0.5, 1])$ the map α is a homeomorphism to its image in P.

We now apply Lemma 1.8 on homotopy triads, using the following notations: For A and B in 1.8, we now have A and P; the map $f : A \to B$ is now $\alpha : A \to P$. For X_1 and X_2 in 1.8, we have $|\epsilon_1|^{-1}([0, 1))$ and $|\epsilon_1|^{-1}((0.5, 1])$. For U_1 and U_2 in 1.8, we have $\alpha(|\epsilon_1|^{-1}([0, 1)))$ and $\alpha(|\epsilon_1|^{-1}((0.5, 1]))$. With these notations, then, the maps $X_2 \to U_2$ and $(X_1 \cap X_2) \to (U_1 \cap U_2)$ given by α are homeomorphisms. Now, X_1 contains K and U_1 contains B, each as deformation retracts, and so from the assumption that $K \to B$ is k-connected, we get the conclusion that $X_1 \to U_1$ is k-connected. Thus, the map $X_1 \to U_1$ is k-connected, and the maps $X_2 \to U_2$ and $(X_1 \cap X_2) \to (U_1 \cap U_2)$ are ∞-connected. It follows then from 1.8 that the map $\alpha : A \to P$ is k-connected.

$\square$

2. Quasi simple filtration

From now on, "space" will mean "polyhedron" and "map" will mean "polyhedral map".

2.1. SIMPLE MAP: That a map $f : A \to B$ is "simple" over $X \subset B$, means that the restriction of f to the inverse image of X is a homeomorphism onto X. (I.e., f yields $f^{-1}(X) \approx X$.)

2.2. QUASI SIMPLE FILTRATION: That a space A is "qsf" ("quasi simply filtrated"), means: A is locally compact, and for every compact $X \subset A$, there exists compact, 1-connected B and a map $f : B \to A$, such that f is simple over X. (The terminology used in a preprint circulated earlier by myself used the terminology "Property B" for "qsf". The definition is due to Stephen G. Brick; his version of the results sketched here is the paper [B].)

A qsf space A can be easily seen to be 1-connected itself; the important words in the definition are "compact" and "1-connected". Thus, that the space A is qsf, means intuitively that A can be approximated by 1-connected and compact pieces; something like the idea of being an increasing union of simply-connected pieces, but we do not require the pieces themselves to be embedded into A out near infinity.

2.3. QSF-COVERED COMPACT POLYHEDRA AND QSF GROUPS: A compact 0-connected space A is said to be "qsf-covered", when its universal cover $\widetilde{A}$ is qsf. We shall then call a finitely presented group G "qsf", if it is the fundamental group of some compact A which is itself qsf-covered. (It follows from results below, 3.5, that if $G = \pi_1(A)$ for some qsf-covered A, then for every compact 0-connected B with $\pi_1(B) \approx G$, we can conclude that B is qsf-covered. Thus, qsf is a group-theoretical property.)

2.4. Theorem. *Suppose that A and B are locally compact. Let $f : A \to B$ be a proper map. Then:*

(a) *If A is qsf and each $f^{-1}(x)$ is 0-connected, then B is qsf.*

(b) *If B is qsf and each $f^{-1}(x)$ is 1-connected, then A is qsf.*

(c) *Thus, if for every $x \in B$, the set $f^{-1}(x)$ is compact and 1-connected, we conclude: A is qsf if and only if B is qsf.*

Proof. Clearly, (c), which is the main point, is a consequence of (a) and (b). We should also remark that "0-connected" is taken to imply "non-empty", so that in this Lemma, the map f is always surjective.

THE PROOF OF (a): Let $X \subset B$ be compact. Since f is proper, the set $Y = f^{-1}(X)$ is compact. Assuming that A is qsf, there is a compact 1-connected T and map $g : T \to A$ which is simple over Y. We identify $g^{-1}(Y)$ with Y. We thus have

$$
\begin{array}{ccc}
f^{-1}(X) = Y & \xrightarrow{\ \subset\ } & T \\
\downarrow{\scriptstyle f} & & \downarrow{\scriptstyle fg} \\
X & \xrightarrow{\ \subset\ } & B
\end{array}
$$

This is the situation of 1.10. There exists compact P containing a copy of X; this P is an approximation to what we would mean by $T/(f^{-1}(X) = X)$. There is a map $P \to B$ in this picture, and the conclusion "$f^{-1}(B) = B$" in 1.10 translates to show that $P \to B$ is simple over X. By assumption, each $f^{-1}(x) \subset Y$ is 0-connected; by 1.9, then, the map $Y \to X$ is 1-connected, and so by 1.10 the map $T \to P$ is 1-connected. Because T is 1-connected, this implies that P is 1-connected. End of proof of (a).

THE PROOF OF (b): Let $X \subset A$ be compact; let $Y = f(X)$, a compact subset of B. Assuming B is qsf, there exists $g : T \to B$ such that T is compact 1-connected and g is simple over Y. Then g and f map T and A to B; T is compact, and f is proper; thus by 1.6, the pullback exists; call the pullback P; there is a map in the pullback diagram $g' : P \to A$, and another map $f' : P \to T$; the map f' has inverse images of points the same as inverse images of points under f, thus each such is 1-connected and compact. Now, 1.9 implies that f' is 2-connected, and so $f'_* : \pi_1(P) \to \pi_1(T)$ is an isomorphism. This shows that P is 1-connected, since T is. We can now examine the map $g' : P \to A$; the fact that g is simple over Y yields the consequence that g' is simple over $f^{-1}(Y)$ which contains X. $\square$

3. Qsf-covered polyhedra and groups

3.1. Lemma.   *Suppose P is a compact connected polyhedron with given triangulation T. Let $Q = |T^{(2)}|$ be its 2-skeleton. Then P is qsf-covered if and only if Q is qsf-covered.*

Proof. By induction on the dimension of P, starting with dimension 2. The inductive step is as follows: Let $n \geq 3$. Let $K = |T^{(n-1)}|$ and $L = |T^{(n)}|$. We have $\pi_1(K) \approx \pi_1(L)$ and, on universal covers, $\widetilde{K} \subset \widetilde{L}$. We show that $\widetilde{K}$ is qsf if and only if $\widetilde{L}$ is qsf.

There is a triangulation of $\widetilde{L}$ lying over the triangulation $T^{(n)}$, whose $(n-1)$-skeleton is a triangulation of $\widetilde{K}$. We call these the "standard triangulations". Suppose that $\widetilde{L}$ is qsf. Let X be a compact subset of $\widetilde{K}$. There is a finite subcomplex S of the standard triangulation of $\widetilde{L}$, such that X lies well within the interior of $|S|$ in $\widetilde{L}$; we can take S to be the closed star of the closed star of any finite subcomplex containing X. Since $\widetilde{L}$ is qsf, there is a compact 1-connected B and map $f : B \to \widetilde{L}$ which is simple over $|S|$; we identify $|S|$ with $f^{-1}(|S|) \subset B$. Then B has a triangulation whose simplexes near X are the simplexes of the standard triangulation, and whose simplexes away from X are fairly small; the map f then has a simplicial approximation $\hat{f}$ from this triangulation of B into the standard triangulation, such that near X it is, so to speak, the identity map. We can also manage to get $\hat{f}^{-1}(\widetilde{L} \setminus X) \subset f^{-1}(\widetilde{L} \setminus X)$, so that $\hat{f}$ is simple over X. Now, remove all the open simplexes of dimension n from $\widetilde{L}$, getting $\widetilde{K}$; from B remove all the open simplexes mapping into these by $\hat{f}$; all such simplexes being removed are of dimension 3 or more, and so the result, $\hat{B}$, is still simply connected, and the restriction of $\hat{f}$ to $\hat{B}$ is simple over X. This shows that $\widetilde{K}$ is qsf.

Conversely, suppose that $\widetilde{K}$ is qsf. Let X be a compact subset of $\widetilde{L}$. Let S be a finite subcomplex of the standard triangulation of $\widetilde{L}$, with $X \subset |S|$, and let $Y = |S| \cap \widetilde{K}$. Since $\widetilde{K}$ is qsf, there exists compact 1-connected B and

$f : B \to \widetilde{K}$ which is simple over Y. Now, we add to B the n-simplexes of S; the result is $\hat{B}$ and an extension $\hat{f}$ of f, to map $\hat{B} \to \widetilde{L}$ simple over $|S|$. We know that $\hat{B}$ is 1-connected, since $n \geq 2$. $\qquad\qquad\qquad\qquad$ □

3.2. REMARK: There is another intuition about the above Lemma 3.1: We could obtain the general case from the 2-dimensional case by adding on high-dimensional cells; adding on a single n-cell can be done in two steps. First we add on a shell of the form $S^{n-1} \times [0,1]$, i.e., creating a mapping cylinder of the attaching map; we can retract this to the previous stage by a map whose inverse images are all points or line segments, and then we can apply 2.4 to the universal cover. Then we identify the sphere $S^{n-1} \times 0$ to a point; this is a thorny matter, since we can't simply "identify" such things but have to use techniques such as 1.10; thus the hypotheses of 2.4 might not, conceivably, be exactly true. Perhaps there is a better way of stating 2.4, so that it could apply to such polyhedral near-identifications.

3.3. Lemma. *If K and L are two compact connected 2-dimensional polyhedra with isomorphic π_1, then there exists a compact polyhedron M and compact subpolyhedra K' and L', such that M collapses onto each of K' and L'; furthermore, K' is the wedge of K and a finite number of S^2's, and similarly L' is the wedge of L and a finite number of S^2's.*

3.4. COLLAPSING: Explicitly, the "collapsing" referred to here involves simplicial structures. One says that A collapses to B, when there is a triangulation of A with B covered by a subcomplex, and there is a sequence of elementary collapses to get from A to B. An elementary collapse from A to B involves some simplex σ of A not in B, and a face τ of σ which is a face of no other simplex of A (one says that τ is a "free face" of A); one then removes the interiors of σ and τ to get B.

This result 3.3 goes back to J.H.C. Whitehead [W]; it is a combination of Whitehead's Theorem 12 on page 266 with his Corollary on page 270. Whitehead's proof involved looking at certain moves changing one group presentation into another of the same group; these are "Reidemeister moves" or "Tietze transformations". Brick's method [B] of proving the next Theorem uses these moves more directly.

3.5. Theorem. *If K and L are two compact connected polyhedra with isomorphic π_1, then K is qsf-covered if and only if L is qsf-covered.*

Proof. By 3.1, we can restrict ourselves to 2-dimensional polyhedra. By 3.3, we need consider only the two cases: (a) L is the wedge of K with a 2-sphere; (b) L collapses to K. In case (a), there is a map from L to K such that the inverse image of every point is either a point or a 2-sphere, and so by 2.4 applied to universal covers, L is qsf-covered iff K is. Case (b) can be reduced, inductively, to one elementary collapse; if L collapses to K by an elementary collapse at the free face τ and simplex σ, then one can subdivide by adding one vertex in τ and retract L to K by sending this vertex to the vertex of σ opposite τ; as a polyhedral map $r : L \to K$, the inverse image of each point is either a singleton or a closed line segment, and thus, 2.4 applies to the lift of r to the universal covers. $\square$

4. Immersions and towers

Note that the definition of qsf (in 2.2) involved maps $B \to A$ which were simple over X. We assumed only that the map $B \to A$ was polyhedral. It turns out that this implies that we can manage to find B and $B \to A$ which is a local embedding ("immersion") that is obtainable by a tower of covering projections and embeddings. It can thus be thickened up to a regular neighborhood situation, so that, for instance, if A is an n-manifold, then B will be an n-manifold as well, and the map $B \to A$ will be an immersion of manifolds. This "tower" idea to obtain an immersion is due to S.M. Gersten (similar results occur in work of Poénaru not using towers; in fact towers have been used in 3-manifolds before by Papakyriakopoulos [Pa] among others, and earlier in the theory of infinite groups by Magnus [Ma]).

4.1. TOWER: Given a map $\phi : B \to A$ of spaces, a "tower of height n" over this map ϕ, consists of the following:

(a) Spaces $B_0, \ldots, B_n$, and $A = C_0, C_1, \ldots, C_n$.
(b) Covering-space projections $p_{i+1} : C_{i+1} \to B_i$. That is, C_{i+1} is a covering space of B_i, and the projection is called p_{i+1}.
(c) Inclusion maps $j_i : B_i \to C_i$.
(d) Surjective maps $\phi_i : B \to B_i$, such that $p_i j_i \phi_i = \phi_{i-1}$.
(e) The tower starts with $B_0 = \phi(B)$, and $\phi_0 : B \to B_0$ being the map ϕ with the range restricted to B_0, and $j_0 : B_0 \to A$ being the inclusion.

In this definition, we shall always have B a compact polyhedron, and so we can consider only the case that A also is compact. The map ϕ can be triangulated. Then B_0 is covered by a subcomplex of the triangulation of A. Each covering $p_{i+1} : C_{i+1} \to B_i$ will be triangulated without changing the given triangulation of B_i. The lift ϕ_i will then be simplicial on the same triangulation of the source B as originally given, and the polyhedron B_i will then be triangulated as a subcomplex of the triangulation of C_i. The polyhedra B_i are compact, and the polyhedra C_i may not be compact.

In the case we are interested in, B is 1-connected, and each of the covering spaces $p_i : C_i \to B_{i-1}$ can be taken to be a universal covering space. If this is a non-trivial covering, the singularities of ϕ_i will be less than the singularities of ϕ_{i-1}; we can make this precise by counting the number of simplexes of all dimensions in the explicit triangulations of B_i; as i increases, the number of simplexes in B_i increases. But the number of simplexes cannot exceed the number of simplexes in B itself. Therefore there will be some finite n for which B_n has no non-trivial covering space; at that point, we say that the tower terminates, with 1-connected B_n.

We can then examine the composition $j_0 p_1 j_1 p_2 j_2 \cdots p_n j_n : B_n \to A$, and have a proof of the following result:

4.2. Theorem. *Let $\phi : B \to A$ be a map which is simple over $X \subset A$, where B is compact and 1-connected. Then there exists compact 1-connected E and map $\psi : E \to A$ simple over X, such that ψ is a composition of a finite number of inclusions and covering projections. Furthermore, if A is a PL manifold, we can take E to be a PL manifold (with boundary) of the same dimension, with ψ still being a composition of inclusions and covering projections.*

Of course, E is the B_n above. We can thicken everything into being a manifold, by taking regular neighborhoods and coverings. The reason that ψ is simple over X is this:

4.3. Lemma. *If $\phi_n : B \to B_n$ is surjective, and $\psi : B_n \to A$ is such that $\psi \phi_n : B \to A$ is simple over X, then ψ is simple over X.*

5. Unions of qsf spaces

One of Brick's interesting results is:

5.1. Theorem. *An amalgamated free product $A *_C B$ has property qsf, if both A and B are qsf and C is finitely generated.*

Sketch of proof. One constructs a space with fundamental group $A *_C B$ by taking two compact spaces K_A and K_B with groups A and B, and joining them together with the cylinder on a finite 1-complex which represents the generators of C mapped into the two respective groups. This resulting space is to be shown to be qsf-covered, by looking at its universal cover. The universal cover is a treelike arrangement of copies of universal covers of K_A and K_B joined along the C covers. We can show this is qsf by considering only a finite number of pieces at a time, and reduce this to a Lemma about the union of two qsf spaces:

5.2. Lemma. *Let M be a space which is the union of two closed subspaces K and L, such that $K \cap L$ is connected. If both K and L are qsf, then M is qsf.*

Proof of Lemma. (Recall that all "spaces" are locally compact polyhedra.) Let X be a compact subspace of M. Then $X \cap (K \cap L)$ is compact and is contained in a compact connected subspace $Y \subset K \cap L$. Let $X' = X \cup Y$. Let $X_K = X' \cap K$ and $X_L = X' \cap L$. Since K is qsf, there is a compact 1-connected B_K and map $f_K : B_K \to K$ which is simple over X_K. Since L is qsf, there is a compact 1-connected B_L and map $f_L : B_L \to L$ which is simple over X_L. Now, the inverse images, under f_K and f_L, of Y can be identified with Y since these maps are simple over Y. Thus, we can take the union of the two B's and identify the copies of Y, getting $B = B_K \cup_Y B_L$ and $f = f_K \cup f_L : B \to M$ (note that the union of two polyhedra over isomorphic subpolyhedra is a case in which the true pushout of polyhedra exists). Then B is compact, being the union of two compact spaces; B is 1-connected, being the union of two 1-connected spaces along a connected subspace; and f, it is easy to see, is simple over X'.   □

An application of this to 3-manifold theory can be imagined thus: Take a 3-manifold, decompose it into canonical pieces, thus making the fundamental group a graph-product of several factors. Apply 5.1 (or its analogue for HNN-extensions). Thus, to show that the 3-manifold itself is qsf-covered, it would be enough to show that each "canonical piece" is qsf-covered.

Here is another application: We know that there exists a finitely presented group with unsolvable word problem. If F denotes the free group on the generators of the presentation involved, and R denotes the subgroup of $F \times F$ generated by the diagonal elements (a, a) for a belonging to the basis of F, and by the elements $(1, r)$, where r ranges over the set of relators, then $R \cap (1 \times F)$ is $1 \times$ (the normal closure in F of the set of relators). Thus, the generalized (relative) word problem for R in $F \times F$ is unsolvable. Hence, taking the "double" amalgamated free product

$$H = (F \times F)_1 *_R (F \times F)_2,$$

we can ask whether, for $u \in F$, the element $(1, u)_1 \cdot (1, u)_2^{-1}$ is the identity element in H. This is equivalent, because of the structure of amalgamated free products, to asking whether u belongs to the normal closure in F of the set of relators. This problem is unsolvable, and so the word problem for H is unsolvable. Note that H has a finite presentation, since the subgroup R is finitely generated. Now, a finitely generated free group F, and its product with itself are easily seen to be qsf (the universal cover of the product of two finite 1-complexes is the increasing union of subcomplexes which are contractible, being the products of two finite trees; rather more general results about products are true, of course); and therefore by 5.1, H itself is qsf:

5.3. Corollary.   *There exists a finitely presented group which is qsf and which has unsolvable word problem.*

(The argument above has some similarities with matters in [Mi], and I have heard that Miller and D. Collins have an even better example, a finitely presented group for which the 2-complex of the presentation is aspherical, such that the group is obtained from a process of a finite number of steps, each of which is an amalgamation or HNN extension over a finitely generated free group, starting with free groups.)

This suggests an embarassing:

5.4. UNSOLVED PROBLEM: Is there a finitely presented group which is not qsf?

6. Universal covers of 3-manifolds

The ideas of this paper apply to (second countable) 3-manifolds because of the theorem that 3-manifolds are triangulable [Mo].

Suppose that M^3 is a closed, aspherical 3-manifold. If $\pi_1(M)$ is qsf, then for any compact connected subset $X \subset \widetilde{M}$, there is a compact, 1-connected B and a map, simple over X, $f : B \to \widetilde{M}$. By the tower technique 4.2 we can take B to be a compact 3-manifold with boundary; the boundary of B, since B is 1-connected, consists of 2-spheres. At least one of these 2-spheres S will be non-bounding in $\widetilde{M} \setminus \{x\}$ for a point $x \in X$. Thus, by the Sphere Theorem [Pa] [He] [Mo], in a neighborhood of $f(S)$ in $\widetilde{M}$, there is an embedded 2-sphere Σ, non-bounding in the complement of x. But Σ bounds a compact, 1-connected submanifold Δ of $\widetilde{M}$, which must contain x, and which therefore must contain all of X, since X is connected and disjoint from Σ. If $\widetilde{M}$ is "irreducible", then Δ must be a 3-cell, and so the whole manifold $\widetilde{M}$ is the union of an increasing sequence of 3-cells, and so is (by M. Brown's theorem [Br]), homeomorphic to $\mathbb{R}^3$. Meeks, Simon, and Yau ("Equivariant Sphere Theorem") ([MSY], [DD], [JR]) have shown that this irreducibility condition follows from assuming that M itself is irreducible (i.e., any tamely embedded 2-sphere in M bounds a topological 3-cell in M). Thus we have:

6.1. Theorem. *Suppose that M is an irreducible, aspherical, closed 3-manifold, and that $\pi_1(M)$ is qsf. Then the universal cover $\widetilde{M}$ is homeomorphic to $\mathbb{R}^3$.*

7. Further problems

How can one describe various classes of groups which are qsf? Some results about this are in [B], [BM], and [MT]. The question in 5.4, as to whether every finitely presented group is qsf, is still unsolved. [Added in Proof: S. T. Tschantz says he has an example of such a non-qsf finitely presented group.] One can conjecture that certain cases, such as fundamental groups of high-dimensional closed aspherical manifolds, are always qsf.

Suppose that $G = \pi_1(K)$, where K is compact and has universal cover $\widetilde{K}$ which is n-connected. We can ask, in analogy to the definition of qsf, whether $\widetilde{K}$ can be approximated by n-connected compact polyhedra. This might lead to a "qsf(n)" property. Conceivably this series of properties and the original qsf property itself might turn out to be significant in group theory.

The technical difficulties in the definitions and arguments in this paper are annoying. My instinct was to utilize CW-complexes and say it was all obvious; but in the usual version of CW theory there are many things that are not true: The image of a subcomplex under a cellular map is not necessarily a subcomplex of the target; the preimage of a subcomplex and more generally the pullback of a CW diagram is not a subcomplex and does not have a CW structure. Identification spaces work out, however, fairly reasonably. In trying to smooth out the CW structures and restrict the class of maps that can be considered, the subject matter seems to be inexorably drawn towards polyhedra.

This topic seems to be purely homotopy-theoretical and so ought to have some description in those terms. Thus, to say that P is qsf, when P is a contractible 3-manifold, is equivalent to saying that P is 1-connected at ∞. In other dimensions, however, qsf is not the same as being 1-connected at ∞.

In another direction, there are non triangulable manifolds and other non polyhedral spaces, for which there might be some sort of qsf theory, with towers and so on. This would require a geometric theory more general than polyhedra; perhaps, absolute neighborhood retracts and maps which have some smoothness property; perhaps, the inverse images of sub-ANR's should be ANR's themselves. In such a theory, there would be interesting problems, such as those related to the tower constructions and why certain towers terminate at a finite stage.

References

[AH] P. Alexandroff and H. Hopf, *Topologie*, Chelsea, 1972. (Originally published Berlin, 1935.)

[B]   S. Brick, *Filtrations of universal covers and a property of groups*, preprint, Univ. of California, Berkeley, 1991.

[BM] S. Brick and M. Mihalik, *The QSF Property for Groups and Spaces*, preprint.

[Br] M. Brown, *The monotone union of open n-cells is an open n-cell*, Proc. Amer. Math. Soc. **12** (1961), pp. 812–814.

[DD] W. Dicks and M. J. Dunwoody, *Groups acting on graphs*, Cambridge Univ. Press, 1989.

[G]   S. M. Gersten, *Isoperimetric and isodiametric functions of finite presentations*, these Proceedings.

[GS]   S. M. Gersten and J. Stallings, *Casson's idea about 3-manifolds whose universal cover is* $\mathbb{R}^3$, International J. of Alg. and Comp. **1** (1991), pp. 395–406.

[He]   J. Hempel, *3-Manifolds*, Annals of Math. Study 86 Princeton Univ. Press, 1976.

[JR]   W. Jaco and J. H. Rubinstein, *PL equivariant surgery and invariant decompositions of 3-manifolds*, Adv. in Math. **73** (1989), pp. 149–191.

[Ma]   W. Magnus, *Über diskontinuierliche Gruppen mit einer definierenden Relation (Der Freiheitssatz)*, J. reine u. angew. Math. **163** (1931), pp. 52–74.

[Mi]   C. F. Miller III, *On Group-Theoretic Decision Problems and Their Classification*, Annals of Math. Study 68, Princeton Univ. Press, 1971.

[Mo]   E. E. Moise, *Geometric Topology in Dimensions 2 and 3*, Graduate Texts in Mathematics 47 Springer, 1977.

[MT]   M. L. Mihalik and S. T. Tschantz, *Tame Combings and the Quasi-simply-filtered Condition for Groups*, to appear.

[MSY]   W. Meeks III, L. Simon, and S-T Yau, *Embedded minimal surfaces, exotic spheres, and manifolds with positive Ricci curvature*, Ann. of Math. **116** (1982), pp. 621–659.

[Pa]   C. D. Papakyriakopoulos, *On Dehn's Lemma and the asphericity of knots*, Ann. of Math. **66** (1957), pp. 1–26.

[P]   V. Poénaru, *Geometry "à la Gromov" for the fundamental group of a closed 3-manifold M^3 and the simple connectivity at infinity of $\widetilde{M}^3$*, preprint, Université de Paris-Sud, Orsay, November 1990.

[W]   J. H. C. Whitehead, *Simplicial spaces, nuclei, and m-groups*, Proc. London Math. Soc. **45** (1939), pp. 243–327.

A Note on Accessibility

G.A. Swarup

Department of Mathematics, University of Melbourne, Melbourne, Australia.

We simplify Dunwoody's proof of accessibility of almost finitely presented groups [3,2]; and as in [4], the proof goes through for groups which act on planar complexes. The main difference between Dunwoody's proof and the proof here is that we do not use the existence theorem for splittings (Theorem 4.1 of [3], and Theorem 5.9 of [2]) to prove accessibility; there is also a proof due to Bestvina and Feighn [1] which does not use the existence of splittings. We recall that the accessibility assumptions arise in topological situations ([6], [7] and [8]) and the question of accessibility was first raised by C.T.C. Wall [8]. Our interest in accessibility came from the extension of Dehn's lemma to higher dimensional knots [7], which itself was a by-product of an attempt to show that higher dimensional knot groups were not free products.

We recall Dunwoody's extension of the theory of normal surfaces to simplicial 2-complexes. If K is a simplicial 2-complex (we use the same symbol for the geometric realization), a *pattern* P in K is a subset of K such that

(1) $P \cap K^0 = \emptyset$,

(2) $P \cap \sigma$ for any closed 2-simplex σ consists of at most a finite number of segments which are disjoint and each of which joins different sides of σ (see Fig. 1.a), and

(3) $P \cap \rho$ for any 1-simplex ρ of K consists of at most a finite number of points.

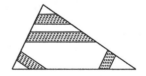

Figure 1.a   Figure 1.b

A connected pattern is called a *track*. A track can be thickened to a *band* (Fig. 1.b, see [2], VI 3). We call a track *2-sided* if the thickening is an *untwisted band.* If the thickening is a twisted band, the track is said to be *1-sided.* If the number of segments in a track is finite or if it is a point, we call the track a *finite track.* Dunwoody extends the Kneser-Haken finiteness technique (see [5]) to show

Theorem 1. (Theorem 2.2 of [3] and Theorem 3.8 of Chapter VI in [2])
For a finite complex K, there is a positive integer $n(K)$ such that, if $t_1, \ldots, t_m$ are disjoint tracks in K and if $m > n(K)$, then at least two of the tracks are parallel (i.e. they bound an untwisted band).

We next observe how tracks arise from splittings of groups. (The following argument seems to be favoured by topologists. Dunwoody suggests an argument modelled after the one on pp. 231–232 of [2]). For the rest of this exposition, we have the following set up:
K is a finite complex, $a : \pi_1(K) \to G$ is a surjection of groups, $\pi : \tilde{K} \to K$ the cover of K corresponding to the kernel of α. We assume that $\tilde{K}$ has the following *planarity property*: the natural map $r : H_c^1(\tilde{K}; \mathbb{Z}_2) \to H^1(\tilde{K}; \mathbb{Z}_2)$ is zero. We identify G as the group of covering translations of π and $\tilde{K}/G = K$.

We want to show that G is accessible. Suppose that G has a non-trivial decomposition as an amalgamated product $G = G_1 *_{G_0} G_2$ or an HNN-extension $G_1 *_{G_0}$ with G_0 finite. We first construct a complex $L = L_1 \cup_{L_0} L_2$ or $L_1 \cup_{L_0}$ with $\pi_1(L_i) = G_i$ and L_0 is 2-sided in L. The last condition means that there is a neighbourhood N of L_0 in L, and N is homeomorphic to $L_0 \times [-1, 1]$ with L_0 corresponding to $L_0 \times \{0\}$. We further assume that $\pi_2(L_i) = 0 = \pi_3(L_i)$, for all i.

Let $f : K \to L$ be a map such that, up to conjugacy, $f_* = \alpha : \pi_1(K) \to \pi_1(L)$, $f(K^0) \subset L - L_0$ and f restricted to any 1-simplex or 2-simplex is transverse to L_0. Thus $f^{-1}(L_0)$ looks almost like a pattern. However, there may be segments with both end points in the same 1-simplex of K. If there are such, starting with an innermost one, we may eliminate them, one by one, as shown in Figure 2.a and 2.b.

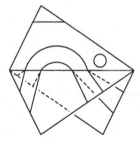

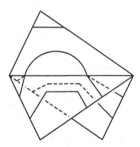

Figure 2.a   Figure 2.b

Thus, after a homotopy, we may assume that $f^{-1}(L_0)$ is a pattern. Suppose that a component s of the pattern $f^{-1}(L_0)$ divides K into two components whose closures are K_1, K_2 and further suppose that $\alpha(\pi_1(K_i)) = \alpha(\pi_1(s))$ for $i = 1$ or 2. Then s can be removed from $f^{-1}(L_0)$ by a further homotopy and so we may assume that none of the components of $f^{-1}(L_0)$ satisfies the above property (which we will temporarily call Property S) which we next formulate in terms of the cover $\pi : \tilde{K} \to K$. Let t be a component of $\pi^{-1}(s)$; t is finite since the stabilizer of t is contained (up to conjugacy) in the finite group G_0.

We next observe that t separates $\tilde{K}$. To see this consider the $\mathbb{Z}_2$-cochain C_t defined by $C_t(y) = |t \cap y|$ mod 2 for every 1-simplex ρ of $\tilde{K}$. If σ is a 2-simplex, t intersects $\partial\sigma$ in an even number of times and therefore $\delta C_t = 0$. Since the image of $H^1_c(\tilde{K}; \mathbb{Z}_2)$ in $H^1(\tilde{K}; \mathbb{Z}_2)$ is zero, C_t is a co-boundary, say $C_t = \delta U = \delta U^*$, where U and U^* are complementary subsets of $\tilde{K}^0$. For each 2-simplex σ of $\tilde{K}$, we colour the components of $\sigma - t$ containing points of U^* black and colour the rest of them alternately (see [2], Chapter VI, Propositions 3.4 and 3.5). It is easy to check that we get a consistent colouring of components $\sigma - t$ which fit together as σ varies to give the two components of $\tilde{K} - t$. Let the closures of the two components be C_1 and C_2.

Property S is equivalent to the assertion that one of C_1 or C_2 is compact. We prove one of the implications, the other is easier. Suppose C_1 is compact, then C_2 is noncompact. Moreover C_1 cannot contain any translate of t since this would contradict the compactness of C_1. A covering translation cannot interchange the sides of t since $\pi(t) = s$ is 2-sided. Hence Stab $t = $ Stab C_1, which implies that s separates K, the closure of one of the components is $\pi(C_1)$ and $\alpha\pi_1(\pi(C_1)) = \alpha\pi_1(\pi(t)) = \alpha\pi_1(s)$. Thus s satisfies property S and should have been eliminated from $f^{-1}(L_0)$. Thus, neither C_1 nor C_2 is compact and C_t represents a non-trivial element of $H^1_c(\tilde{K}; \mathbb{Z}_2)$.

Now consider $\tilde{K} - \pi^{-1}(s) = \tilde{K} - \bigcup_{g \in G} g(t)$. Taking one vertex for each component, and joining them by a segment when the closures intersect, we obtain a tree Γ on which G acts without inversions, since s is 2-sided. This gives us a new decomposition of G. If K is split along s to form $K_1 \cup_s K_2$ or $K_1 \cup_s s$ and $\alpha(\pi_1(K_i)) = H_i$, $\alpha(\pi_1(s)) = H_0$, then $G = H_1 *_{H_0} H_2$ or $H_1 *_{H_0}$ is the decomposition of G given by the above action on the tree Γ.

We next observe that K_i (one of them may be empty, in which case assume $K_2 = \varnothing$,) obtained by splitting K along s has a special type of cell-structure, which we call a *marked cell structure*. (Dunwoody uses similar constructions in [2], p. 264). For this structure, the cells are of six different types as in Figure 3, where the marked cells are dotted.

We are considering only the marked cell complexes which are obtained from simplicial complexes by splitting successively along tracks which for marked cell complexes are assumed to be disjoint from marked cells. Such a track

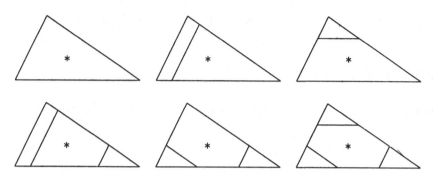

Figure 3

is automatically a track in the original simplicial complex. It is also easy to check that splitting a marked cell complex along a track again yields a marked complex. After the above decomposition $G = H_1 *_{H_0} H_2$ or $H_1 *_{H_0}$ using the tracks, suppose that one of the H_1 is further decomposable as $H_1 = H_1' *_{H_0'} H_2'$ or $H_1' *_{H_0'}$, and continue. At each stage the new track s_i is not parallel to any of the previous tracks (since their fundamental groups map into finite groups) as long as we can decompose any of the new groups that are obtained. By Theorem 1, the process has to stop after a finite number of steps since we cannot find more than $n(K)$ tracks, no two of which are parallel.

References

[1] M. Bestvina and M. Feighn, *Bounding the complexity of simplicial group actions on trees*, Inv. Math. **103** (1991), pp. 449–469.

[2] W. Dicks and M. Dunwoody, *Groups acting on Graphs*, Cambridge University Press, 1990.

[3] M. Dunwoody, *The accessibility of finitely presented groups*, Inv. Math. **81** (1985), pp. 449–457.

[4] J.R.J. Groves and G.A. Swarup, *Remarks on a technique of Dunwoody*, Jour. of Pure and Appl. Algebra **75** (1991), pp. 259–269.

[5] J. Hempel, *3-manifolds*, Annals of Maths Studies 86, Princeton University Press, 1976.

[6] G. Martin and R. Skora, *Group actions on the 2-sphere*, Amer. Jour. Math. **111** (1989), pp. 387–402.

[7] G.A. Swarup, *An Unknotting Criterion*, J. of Pure and Applied Algebra **6** (1975), pp. 291–296.

[8] C.T.C. Wall, *Pairs of relative cohomological dimension one*, J. Pure and Appl. Algebra **1** (1971), pp. 141–154.

Geometric Group Theory 1991 Problem List

Editor: Graham A. Niblo

Mathematics Department, Southampton University, Southampton, SO9 5NH.

Problem 1. [J.M. Alonso] Do torsion free combable groups always have finite cohomological dimension?

The Rips complex shows that the answer is yes for hyperbolic groups, and that combable groups are finitely presented of type FP_∞ (cf. Alonso, Combings of groups). Brown and Geoghegan showed that the Thompson group Γ (see Probl. 13) is torsion free, of type FP_∞ and has infinite cohomological dimension. Gersten has proved that Γ is not combable, since it has an exponential isoperimetric inequality.

Problem 2. [Anon.] Let G be an automatic group. Is there a finite dimensional, contractible CW complex on which G acts properly discontinuously?

Problem 3. [Anon.] Does every negatively curved group contain a proper subgroup of finite index?

Problem 4. [M. Bestvina] Let K be a compact, non-simply connected, acyclic simplicial complex in which the triangulation is full, and let V be the set $\{v_\sigma \mid \sigma$ is a simplex of $K\}$. Define a Coxeter group Γ by the following presentation:

$$\Gamma = \langle V \mid v_\sigma^2 = 1, (v_\sigma v_\tau)^{m(\sigma,\tau)} = 1\rangle,$$

where

$$m(\sigma,\tau) = \begin{cases} 2, & \text{if } \sigma \text{ is a face of } \tau \text{ or vice versa;} \\ \infty, & \text{otherwise.} \end{cases}$$

Then $\mathrm{vcd}\,(\Gamma) = 2$. Is there a finite index subgroup $\Gamma_0 \subset \Gamma$ which admits a 2-dimensional $K(\Gamma_0, 1)$?

Problem 5. [M. Bestvina] *Definition:* An automorphism ϕ of a free group F is a *generalised Dehn twist* with respect to a basis $\{a_1, a_2, \ldots, a_n\}$ for F, if some power ϕ^k has the form:

$$\phi^k = \begin{cases} a_1 \longrightarrow a_1, \\ a_2 \longrightarrow w_2 a_2 v_2, \\ \vdots \\ a_i \longrightarrow w_i a_i v_i, \\ \vdots \\ a_n \longrightarrow w_n a_n v_n, \end{cases}$$

where each w_i and v_i is an element of the subgroup $\langle a_1, \ldots, a_{i-1} \rangle$.

Suppose that G is a finitely generated subgroup of $\operatorname{Aut}(F_n)$ which consists entirely of generalised Dehn twists, each with respect to its own basis. Is there always a basis $\{a_1, a_2, \ldots, a_n\}$ for F and a finite index subgroup of G which consists entirely of generalised Dehn twists with respect to this basis?

Problem 6. [H. Chaltin] Let R be an integral domain, and G be a group for which the group ring RG is also an integral domain. Is DG an integral domain for every integral domain D?

Problem 7. [R. Charney] Are Artin groups of infinite type automatic, bi-automatic or semi-hyperbolic?

Problem 8. [D. Epstein] Is there any connection between isoperimetric inequalities and time estimates for algorithms solving the word problem?

Problem 9. [D. Epstein] Is the class of automatic groups closed under quasi-isometry? Thurston thinks not. Gromov pointed out that this is not even known for co-compact lattices in Lie groups.

Problem 10. [R.G. Fenn] Let R be a finite rack. When is the rack homomorphism $R \rightarrow \operatorname{As}(R)$ injective?

Problem 11. [R.G. Fenn] Give a list of all finite simple racks.

Problem 12. [S. Gersten] Let ϕ be an automorphism of a free group F, and let Γ be the corresponding semidirect product $\Gamma = F \times_\phi \mathbb{Z}$. Is Γ always bi-automatic?

Bestvina and Feighn have shown that if Γ has no subgroups isomorphic to $\mathbb{Z} \times \mathbb{Z}$ then it is hyperbolic, so the answer in this case is yes. On the other hand Gersten showed that if ϕ has the form:

$$\phi = \begin{cases} a \longrightarrow a, \\ b \longrightarrow ba, \\ c \longrightarrow cab, \end{cases}$$

then Γ cannot exist as a subgroup of any group which acts co-compactly on a non-positively curved space. This seems to be evidence in the opposite direction.

Problem 13. [R. Geoghegan] Let F be the group given by the following presentation:

$$\langle x_0, x_1, \ldots, \mid x_i^{-1} x_n x_i = x_{n+1} \ \forall i < n \rangle.$$

It was discovered by R. Thompson, and has been studied by many people (e.g. Freyd-Heller, Hastings-Heller, Dydack, Brown-Geoghegan). It is torsion free and has a finite presentation.

In 1979 Geoghegan conjectured that

a) F has no non-cyclic free subgroups, and

b) F is not amenable.

There are no known examples of a finitely presented group which satisfies both of these properties, but Grigorchuk found a finitely generated group satisfying them both in 1980. Brin and Squiers showed in about 1984 that the Thompson group satisfies a). Show that it satisfies b).

Problem 14. [A. Haefliger] Let Γ be a group acting properly discontinuously and cocompactly on a simply connected Riemannian manifold. Show that Γ contains some element which has no fixed points.

This is connected with the question of whether or not there exists an infinite, finitely presented, torsion group.

Problem 15. [P. de la Harpe] A group Γ of homeomorphisms of S^1 is said to be cyclically n-transitive if, given any two cyclically ordered n-tuples $(x_1, \ldots, x_n)$ and $(y_1, \ldots, y_n)$ there exists $\gamma \in \Gamma$ such that $(x_1, \ldots, x_n) = (\gamma(y_1), \ldots, \gamma(y_n))$. For example $PSL_2(\mathbb{R})$ is a cyclically 3-transitive group and Homeo(S^1) is cyclically n-transitive for all n. Find a group Γ which is cyclically n-transitive but not cyclically $n+1$-transitive for some $n \geq 4$.

Problem 16. [P. de la Harpe] A homeomorphism ϕ of a compact metric space X is said to be hyperbolic if it has exactly two fixed points α and β, and for any neighbourhoods A of α and B of β there is a positive integer N such that for all n larger than N, $\phi^n(X \setminus A) \subset B$ and $\phi^{-n}(X \setminus B) \subset A$. ($X$ is assumed to have more than 3 points)

Let Γ be any lattice in a real, simple Lie group of real rank greater than or equal to 2, e.g. $\Gamma = SL_3(\mathbb{Z})$. Does there exist a compact metric space X and an action of Γ on X which has hyperbolic elements?

Problem 17. [J. Howie, R. Thomas] Is the group $G_{13,4}$ given by the presentation $\langle a, b \mid a^2 = b^3 = (ab)^{13} = [a, b]^4 = 1 \rangle$ finite or infinite? If the exponents 13 and 4 are replaced by any other pair of integers the answer is known.

The group G has a homomorphic image $H \times PSL(2, 25)$, where H is an extension of $PSL(3, 3)$ by $\mathbb{Z}_2^{12}$. This image is too big for existing packages to examine the kernel successfully. On the other hand small cancellation methods which can be used to show that $G_{p,q}$ is infinite for some other values of p and q, fail in this case.

Problem 18. [P.H. Kropholler] Let G be the fundamental group of a closed hyperbolic 3-manifold, and H and K be finitely generated, geometrically finite subgroups of infinite index in G. Do there exist subgroups H_0, K_0 of finite index in H, K respectively such that $\langle H_0, K_0 \rangle$ has infinite index in G?

Problem 19. [P.H. Kropholler] Let G be the fundamental group of a hyperbolic 3-manifold, and let H be a locally free subgroup of G. Is H necessarily free?

Problem 20. [P.H. Kropholler] Mess has shown that the Baumslag-Solitar groups $\langle x, t \mid (x^p)^t = x^q \rangle$ can be embedded in Poincaré duality groups of dimension 5 (PD^5-groups). Can they be embedded in PD^3-groups for $p \neq \pm q$?

Problem 21. [M. Lustig, Y. Minski] Let G be a group acting freely on some $\mathbb{R}$-tree T; let $\|g\|$ denote the hyperbolic length function of g with respect to this action. Fix a generating system S of G, and denote by $|g|_S$ the word length of g with respect to S. The G-action is *thin* if there is a sequence $g_1, g_2, \ldots, \in G \smallsetminus \{1\}$ with $|g_i|_S \cdot \|g_i\| \overset{i \to \infty}{\longrightarrow} 0$, otherwise call the action *thick*. Is the set of thick actions uncountable? Is the set of thick actions of measure zero in some reasonable sense? These are both true if G is free abelian, according to the Falmer pub research group. If F is a free group and $\phi \in \mathrm{Out}\,(F)$ commutes with a homothety of T (i.e. $[\|\cdot\|_T] \in SLF\,(F_n)$ is fixed by ϕ), does it follow that T is thick? This is true if G is the fundamental group of a 2-manifold by work of Minski.

Problem 22. [Y. Minski] Let S be a surface, and L be a geodesic in the Teichmüller space of S. L is said to be *thick* if there is a uniform lower bound on the length of non-peripheral essential curves in S, for all hyperbolic metrics represented by points along L. Are there uncountably many bi-infinite thick geodesics? (H. Masur has shown that the subspace of thick geodesics has zero measure.)
Note that all rays whose endpoints are fixed points of pseudo-Anosov diffeomorphisms are thick. There are "periodic" examples; there should be aperiodic ones as well.

Problem 23. [Y. Moriah] Let G be a group generated by X. An automorphism ϕ of G is said to be *tame* if it is induced by an automorphism of the free group on X. Is there a way to distinguish tame and non-tame automorphisms of G by examining the Cayley graph with respect to X?

Problem 24. [W. Neumann] The *abstract commensurator* of a group Γ is

$$\mathrm{Comm}\,(\Gamma) = \{\text{isomorphisms between finite index subgroups of } \Gamma\}/\sim \,,$$

where $\phi_1 \sim \phi_2$ means that ϕ_1 and ϕ_2 agree on some finite index subgroup. Suppose that Γ is negatively curved, and no finite index subgroup is reducible. Suppose, moreover, that $\mathrm{Comm}\,(\Gamma)$ is "large" in a suitable sense, (maybe $|\mathrm{Comm}\,(\Gamma) : \Gamma| = \infty$ is sufficient?). Is Γ then an arithmetic subgroup of an algebraic group?

Problem 25. [W. Neumann] Let SC denote the abstract commensurator of $\pi_1(F)$ where F is any closed surface of genus greater than one. SC acts effectively on the circle at infinity of $\pi_1(F)$, and is therefore a subgroup of Homeo(S^1). Is it a dense subgroup? Bass showed that the analogue for free groups is true; the abstract commensurator of a free group Γ is dense in Homeo$(\mathcal{C})$ where $\mathcal{C}$ is the Cantor set at infinity of Γ.

Problem 26. [W. Neumann, A Reid] Let G be the group of Isometries of $\mathbb{H}^3$, let K be a knot in S^3 and let $\Gamma < G$ be such that $\mathbb{H}^3/\Gamma$ is the complement of K in S^3. Suppose that Comm$_G(\Gamma) \neq N(\Gamma)$, where

$$\text{Comm}_G(\Gamma) = \{g \in G \mid \Gamma^g \cap \Gamma \text{ has finite index in both } \Gamma \text{ and } \Gamma^g\}$$

and $N(\Gamma)$ is the normaliser of Γ in G. We then conjecture that K is either the figure-8 knot, (the only knot with arithmetic complement), or one of the two Aitcheson-Rubinstein dodecahedral knots. (These both have Comm$_G(\Gamma)$ equal to the isometry group of a regular, ideal dodecahedral tesselation of $\mathbb{H}^3$, and $S^3 \setminus K = \mathbb{H}^3/\Gamma$.)

Problem 27. [F. Paulin] Let G be a negatively curved group. When is ∂G locally connected? If ∂G has local cut points does this imply that G splits over a virtually cyclic subgroup?

Problem 28. [L. Potyagailo] Let F and G be any finitely generated Kleinian subgroups of $SO(1,4)$, where F is an infinite index normal subgroup of G and F, G and G/F all act on a connected subdomain $\Omega \subset S^3$. Show that $\pi_1(\Omega/F)$ is finitely generated if and only if Ω is simply connected. This is true when G is the fundamental group of a closed hyperbolic 3-manifold.

Problem 29. [A. Reid] Let M be a hyperbolic knot complement. Can M contain a separating, totally geodesic surface?

Problem 30. [E. Rips] Is every (torsion free) negatively curved group Hopfian? Note that if the group has no non-trivial actions on any $\mathbb{R}$-tree then this is true.

Problem 31. [E. Rips] Is every freely indecomposable (torsion free) negatively curved group co-Hopfian? This is known when the group has no splitting over a virtually cyclic group.

Problem 32. [H. Short] Gromov and Olshanskii have shown that if a group G has a subquadratic isoperimetric inequality, then it has a linear one. Are there intermediate IPE's, say between n and $n+1$?

Problem 33. [Swarup] Define *quasi-simply filtrated* algebraically. Is there a connection between this notion and the end invariants of a group?

Problem 34. [Swarup] Is there an algebraic annulus theorem? I.e., given an infinite cyclic subgroup $C < G$ such that $e(G, C) = 2$ give reasonable sufficient conditions for G to split over a subgroup commensurable with C.

Printed in the United States
By Bookmasters